10/10

You Are Not a Gadget

you are not a

A MANIFESTO

JARON

LANIER

ALFRED A. KNOPF
NEW YORK
2010

THIS IS A BORZOI BOOK PUBLISHED BY ALFRED A. KNOPF

Copyright © 2010 by Jaron Lanier

Grateful acknowledgment is made to Imprint Academic for permission to reprint material
by Jaron Lanier that was originally published in the *Journal of Consciousness Studies*.

Portions of this work also originally appeared in *Discover, Think Magazine*, and on
www.edge.org.

Library of Congress Cataloging-in-Publication Data
Lanier, Jaron.
    You are not a gadget / by Jaron Lanier.—1st ed.
    p. cm.
    ISBN 978-0-307-26964-5 (hbk.)
    1. Information technology—Social aspects.   2. Technological
    innovations—Social aspects.   3. Technology—Social aspects.   I. Title.
HM851.L358 2010
303.48'33—dc22   2009020298

Manufactured in the United States of America
First Edition

This book is dedicated
to my friends and colleagues in the digital revolution.
Thank you for considering my challenges constructively,
as they are intended.

Thanks to Lilly for giving me yearning,
and Ellery for giving me eccentricity,
to Lena for the mrping,
and to Lilibell, for teaching me to read anew.

contents

PReFaCe

IT'S EARLY in the twenty-first century, and that means that these words will mostly be read by nonpersons—automatons or numb mobs composed of people who are no longer acting as individuals. The words will be minced into atomized search-engine keywords within industrial cloud computing facilities located in remote, often secret locations around the world. They will be copied millions of times by algorithms designed to send an advertisement to some person somewhere who happens to resonate with some fragment of what I say. They will be scanned, rehashed, and misrepresented by crowds of quick and sloppy readers into wikis and automatically aggregated wireless text message streams.

Reactions will repeatedly degenerate into mindless chains of anonymous insults and inarticulate controversies. Algorithms will find correlations between those who read my words and their purchases, their romantic adventures, their debts, and, soon, their genes. Ultimately these words will contribute to the fortunes of those few who have been able to position themselves as lords of the computing clouds.

The vast fanning out of the fates of these words will take place almost entirely in the lifeless world of pure information. Real human eyes will read these words in only a tiny minority of the cases.

And yet it is you, the person, the rarity among my readers, I hope to reach.

The words in this book are written for people, not computers.

I want to say: You have to be somebody before you can share yourself.

PART ONE

WHAT IS A PERSON?

MISSING PERSONS

SOFTWARE EXPRESSES IDEAS about everything from the nature of a musical note to the nature of personhood. Software is also subject to an exceptionally rigid process of "lock-in." Therefore, ideas (in the present era, when human affairs are increasingly software driven) have become more subject to lock-in than in previous eras. Most of the ideas that have been locked in so far are not so bad, but some of the so-called web 2.0 ideas are stinkers, so we ought to reject them while we still can.

Speech is the mirror of the soul; as a man speaks, so is he.

PUBLILIUS SYRUS

Fragments Are Not People

Something started to go wrong with the digital revolution around the turn of the twenty-first century. The World Wide Web was flooded by a torrent of petty designs sometimes called web 2.0. This ideology promotes radical freedom on the surface of the web, but that freedom, ironically, is more for machines than people. Nevertheless, it is sometimes referred to as "open culture."

Anonymous blog comments, vapid video pranks, and lightweight mashups may seem trivial and harmless, but as a whole, this widespread practice of fragmentary, impersonal communication has demeaned interpersonal interaction.

Communication is now often experienced as a superhuman phenomenon that towers above individuals. A new generation has come of age with a reduced expectation of what a person can be, and of who each person might become.

The Most Important Thing About a Technology Is How It Changes People

When I work with experimental digital gadgets, like new variations on virtual reality, in a lab environment, I am always reminded of how small changes in the details of a digital design can have profound unforeseen effects on the experiences of the humans who are playing with it. The slightest change in something as seemingly trivial as the ease of use of a button can sometimes completely alter behavior patterns.

For instance, Stanford University researcher Jeremy Bailenson has demonstrated that changing the height of one's avatar in immersive virtual reality transforms self-esteem and social self-perception. Technologies are extensions of ourselves, and, like the avatars in Jeremy's lab, our identities can be shifted by the quirks of gadgets. It is impossible to work with information technology without also engaging in social engineering.

One might ask, "If I am blogging, twittering, and wikiing a lot, how does that change who I am?" or "If the 'hive mind' is my audience, who am I?" We inventors of digital technologies are like stand-up comedians or neurosurgeons, in that our work resonates with deep philosophical questions; unfortunately, we've proven to be poor philosophers lately.

When developers of digital technologies design a program that requires you to interact with a computer as if it were a person, they ask you to accept in some corner of your brain that you might also be conceived of as a program. When they design an internet service that is edited by a vast anonymous crowd, they are suggesting that a random crowd of humans is an organism with a legitimate point of view.

Different media designs stimulate different potentials in human nature. We shouldn't seek to make the pack mentality as efficient as possible. We should instead seek to inspire the phenomenon of individual intelligence.

"What is a person?" If I knew the answer to that, I might be able to program an artificial person in a computer. But I can't. Being a person is not a pat formula, but a quest, a mystery, a leap of faith.

Optimism

It would be hard for anyone, let alone a technologist, to get up in the morning without the faith that the future can be better than the past.

Back in the 1980s, when the internet was only available to small number of pioneers, I was often confronted by people who feared that the strange technologies I was working on, like virtual reality, might unleash the demons of human nature. For instance, would people become addicted to virtual reality as if it were a drug? Would they become trapped in it, unable to escape back to the physical world where the rest of us live? Some of the questions were silly, and others were prescient.

How Politics Influences Information Technology

I was part of a merry band of idealists back then. If you had dropped in on, say, me and John Perry Barlow, who would become a cofounder of the Electronic Frontier Foundation, or Kevin Kelly, who would become the founding editor of *Wired* magazine, for lunch in the 1980s, these are the sorts of ideas we were bouncing around and arguing about. Ideals are important in the world of technology, but the mechanism by which ideals influence events is different than in other spheres of life. Technologists don't use persuasion to influence you—or, at least, we don't do it very well. There are a few master communicators among us (like Steve Jobs), but for the most part we aren't particularly seductive.

We make up extensions to your being, like remote eyes and ears (webcams and mobile phones) and expanded memory (the world of details you can search for online). These become the structures by which you connect to the world and other people. These structures in turn can change how you conceive of yourself and the world. We tinker with your philosophy by direct manipulation of your cognitive experience, not indirectly, through argument. It takes only a tiny group of engineers to create technology that can shape the entire future of human experience with incredible speed. Therefore, crucial arguments about the human relationship with technology should take place between developers and users before such direct manipulations are designed. This book is about those arguments.

The design of the web as it appears today was not inevitable. In the early 1990s, there were perhaps dozens of credible efforts to come up with a design for presenting networked digital information in a way that would attract more popular use. Companies like General Magic and Xanadu developed alternative designs with fundamentally different qualities that never got out the door.

A single person, Tim Berners-Lee, came to invent the particular design of today's web. The web as it was introduced was minimalist, in that it assumed just about as little as possible about what a web page would be like. It was also open, in that no page was preferred by the architecture over another, and all pages were accessible to all. It also emphasized responsibility, because only the owner of a website was able to make sure that their site was available to be visited.

Berners-Lee's initial motivation was to serve a community of physicists, not the whole world. Even so, the atmosphere in which the design of the web was embraced by early adopters was influenced by idealistic discussions. In the period before the web was born, the ideas in play were radically optimistic and gained traction in the community, and then in the world at large.

Since we make up so much from scratch when we build information technologies, how do we think about which ones are best? With the kind of radical freedom we find in digital systems comes a disorienting moral challenge. We make it all up—so what shall we make up? Alas, that dilemma—of having so much freedom—is chimerical.

As a program grows in size and complexity, the software can become a cruel maze. When other programmers get involved, it can feel like a labyrinth. If you are clever enough, you can write any small program from scratch, but it takes a huge amount of effort (and more than a little luck) to successfully modify a large program, especially if other programs are already depending on it. Even the best software development groups periodically find themselves caught in a swarm of bugs and design conundrums.

Little programs are delightful to write in isolation, but the process of maintaining large-scale software is always miserable. Because of this, digital technology tempts the programmer's psyche into a kind of schizophrenia. There is constant confusion between real and ideal computers. Technologists wish every program behaved like a brand-new, playful little program, and will use any available psychological strategy to avoid thinking about computers realistically.

The brittle character of maturing computer programs can cause digital designs to get frozen into place by a process known as lock-in. This happens when many software programs are designed to work with an existing one. The process of significantly changing software in a situation in which a lot of other software is dependent on it is the hardest thing to do. So it almost never happens.

Occasionally, a Digital Eden Appears

One day in the early 1980s, a music synthesizer designer named Dave Smith casually made up a way to represent musical notes. It was called MIDI. His approach conceived of music from a keyboard player's point of view. MIDI was made of digital patterns that represented keyboard events like "key-down" and "key-up."

That meant it could not describe the curvy, transient expressions a singer or a saxophone player can produce. It could only describe the tile mosaic world of the keyboardist, not the watercolor world of the violin. But there was no reason for MIDI to be concerned with the whole of musical expression, since Dave only wanted to connect some synthesizers together so that he could have a larger palette of sounds while playing a single keyboard.

In spite of its limitations, MIDI became the standard scheme to represent music in software. Music programs and synthesizers were designed to work with it, and it quickly proved impractical to change or dispose of all that software and hardware. MIDI became entrenched, and despite Herculean efforts to reform it on many occasions by a multidecade-long parade of powerful international commercial, academic, and professional organizations, it remains so.

Standards and their inevitable lack of prescience posed a nuisance before computers, of course. Railroad gauges—the dimensions of the tracks—are one example. The London Tube was designed with narrow tracks and matching tunnels that, on several of the lines, cannot accommodate air-conditioning, because there is no room to ventilate the hot air from the trains. Thus, tens of thousands of modern-day residents in one of the world's richest cities must suffer a stifling commute because of an inflexible design decision made more than one hundred years ago.

But software is worse than railroads, because it must always adhere with absolute perfection to a boundlessly particular, arbitrary, tangled, intractable messiness. The engineering requirements are so stringent and perverse that adapting to shifting standards can be an endless struggle. So while lock-in may be a gangster in the world of railroads, it is an absolute tyrant in the digital world.

Life on the Curved Surface of Moore's Law

The fateful, unnerving aspect of information technology is that a particular design will occasionally happen to fill a niche and, once implemented, turn out to be unalterable. It becomes a permanent fixture from then on, even though a better design might just as well have taken its place before the moment of entrenchment. A mere annoyance then explodes into a cataclysmic challenge because the raw power of computers grows exponentially. In the world of computers, this is known as Moore's law.

Computers have gotten *millions* of times more powerful, and immensely more common and more connected, since my career began—which was not so very long ago. It's as if you kneel to plant a seed of a tree and it grows so fast that it swallows your whole village before you can even rise to your feet.

So software presents what often feels like an unfair level of responsibility to technologists. Because computers are growing more powerful at an exponential rate, the designers and programmers of technology must be extremely careful when they make design choices. The consequences of tiny, initially inconsequential decisions often are amplified to become defining, unchangeable rules of our lives.

MIDI now exists in your phone and in billions of other devices. It is the lattice on which almost all the popular music you hear is built. Much of the sound around us—the ambient music and audio beeps, the ring-tones and alarms—are conceived in MIDI. The whole of the human auditory experience has become filled with discrete notes that fit in a grid.

Someday a digital design for describing speech, allowing computers to sound better than they do now when they speak to us, will get locked in. That design might then be adapted to music, and perhaps a more fluid and expressive sort of digital music will be developed. But even if that happens, a thousand years from now, when a descendant of ours is traveling at relativistic speeds to explore a new star system, she will prob-ably be annoyed by some awful beepy MIDI-driven music to alert her that the antimatter filter needs to be recalibrated.

Lock-in Turns Thoughts into Facts

Before MIDI, a musical note was a bottomless idea that transcended absolute definition. It was a way for a musician to think, or a way to teach and document music. It was a mental tool distinguishable from the music itself. Different people could make transcriptions of the same musical recording, for instance, and come up with slightly different scores.

After MIDI, a musical note was no longer just an idea, but a rigid, mandatory structure you couldn't avoid in the aspects of life that had gone digital. The process of lock-in is like a wave gradually washing over the rulebook of life, culling the ambiguities of flexible thoughts as more and more thought structures are solidified into effectively permanent reality.

We can compare lock-in to scientific method. The philosopher Karl Popper was correct when he claimed that science is a process that dis-qualifies thoughts as it proceeds—one can, for example, no longer rea-sonably believe in a flat Earth that sprang into being some thousands of years ago. Science removes ideas from play empirically, for good reason.

Lock-in, however, removes design options based on what is easiest to program, what is politically feasible, what is fashionable, or what is created by chance.

Lock-in removes ideas that do not fit into the winning digital representation scheme, but it also reduces or narrows the ideas it immortalizes, by cutting away the unfathomable penumbra of meaning that distinguishes a word in natural language from a command in a computer program.

The criteria that guide science might be more admirable than those that guide lock-in, but unless we come up with an entirely different way to make software, further lock-ins are guaranteed. Scientific progress, by contrast, always requires determination and can stall because of politics or lack of funding or curiosity. An interesting challenge presents itself: How can a musician cherish the broader, less-defined concept of a note that preceded MIDI, while using MIDI all day long and interacting with other musicians through the filter of MIDI? Is it even worth trying? Should a digital artist just give in to lock-in and accept the infinitely explicit, finite idea of a MIDI note?

If it's important to find the edge of mystery, to ponder the things that can't quite be defined—or rendered into a digital standard—then we will have to perpetually seek out entirely new ideas and objects, abandoning old ones like musical notes. Throughout this book, I'll explore whether people are becoming like MIDI notes—overly defined, and restricted in practice to what can be represented in a computer. This has enormous implications: we can conceivably abandon musical notes, but we can't abandon ourselves.

When Dave made MIDI, I was thrilled. Some friends of mine from the original Macintosh team quickly built a hardware interface so a Mac could use MIDI to control a synthesizer, and I worked up a quick music creation program. We felt so free—but we should have been more thoughtful.

By now, MIDI has become too hard to change, so the culture has changed to make it seem fuller than it was initially intended to be. We have narrowed what we expect from the most commonplace forms of musical sound in order to make the technology adequate. It wasn't Dave's fault. How could he have known?

Digital Reification: Lock-in Turns Philosophy into Reality

A lot of the locked-in ideas about how software is put together come from an old operating system called UNIX. It has some characteristics that are related to MIDI.

While MIDI squeezes musical expression through a limiting model of the actions of keys on a musical keyboard, UNIX does the same for all computation, but using the actions of keys on typewriter-like keyboards. A UNIX program is often similar to a simulation of a person typing quickly.

There's a core design feature in UNIX called a "command line interface." In this system, you type instructions, you hit "return," and the instructions are carried out.[*] A unifying design principle of UNIX is that a program can't tell if a person hit return or a program did so. Since real people are slower than simulated people at operating keyboards, the importance of precise timing is suppressed by this particular idea. As a result, UNIX is based on discrete events that don't have to happen at a precise moment in time. The human organism, meanwhile, is based on continuous sensory, cognitive, and motor processes that have to be synchronized precisely in time. (MIDI falls somewhere in between the concept of time embodied in UNIX and in the human body, being based on discrete events that happen at particular times.)

UNIX expresses too large a belief in discrete abstract symbols and not enough of a belief in temporal, continuous, nonabstract reality; it is more like a typewriter than a dance partner. (Perhaps typewriters or word processors ought to always be instantly responsive, like a dance partner—but that is not yet the case.) UNIX tends to "want" to connect to reality as if reality were a network of fast typists.

If you hope for computers to be designed to serve embodied people as well as possible people, UNIX would have to be considered a bad design. I discovered this in the 1970s, when I tried to make responsive musical

---

[*]The style of UNIX commands has, incredibly, become part of pop culture. For instance, the URLs (universal resource locators) that we use to find web pages these days, like http://www.jaronlanier.com/, are examples of the kind of key press sequences that are ubiquitous in UNIX.

instruments with it. I was trying to do what MIDI does not, which is work with fluid, hard-to-notate aspects of music, and discovered that the underlying philosophy of UNIX was too brittle and clumsy for that.

The arguments in favor of UNIX focused on how computers would get literally millions of times faster in the coming decades. The thinking was that the speed increase would overwhelm the timing problems I was worried about. Indeed, today's computers are millions of times faster, and UNIX has become an ambient part of life. There are some reasonably expressive tools that have UNIX in them, so the speed increase has sufficed to compensate for UNIX's problems in some cases. But not all.

I have an iPhone in my pocket, and sure enough, the thing has what is essentially UNIX in it. An unnerving element of this gadget is that it is haunted by a weird set of unpredictable user interface delays. One's mind waits for the response to the press of a virtual button, but it doesn't come for a while. An odd tension builds during that moment, and easy intuition is replaced by nervousness. It is the ghost of UNIX, still refusing to accommodate the rhythms of my body and my mind, after all these years.

I'm not picking in particular on the iPhone (which I'll praise in another context later on). I could just as easily have chosen any contemporary personal computer. Windows isn't UNIX, but it does share UNIX's idea that a symbol is more important than the flow of time and the underlying continuity of experience.

The grudging relationship between UNIX and the temporal world in which the human body moves and the human mind thinks is a disappointing example of lock-in, but not a disastrous one. Maybe it will even help make it easier for people to appreciate the old-fashioned physical world, as virtual reality gets better. If so, it will have turned out to be a blessing in disguise.

Entrenched Software Philosophies Become Invisible Through Ubiquity

An even deeper locked-in idea is the notion of the file. Once upon a time, not too long ago, plenty of computer scientists thought the idea of the file was not so great.

The first design for something like the World Wide Web, Ted Nelson's Xanadu, conceived of one giant, global file, for instance. The first iteration of the Macintosh, which never shipped, didn't have files. Instead, the whole of a user's productivity accumulated in one big structure, sort of like a singular personal web page. Steve Jobs took the Mac project over from the fellow who started it, the late Jef Raskin, and soon files appeared.

UNIX had files; the Mac as it shipped had files; Windows had files. Files are now part of life; we teach the idea of a file to computer science students as if it were part of nature. In fact, our conception of files may be more persistent than our ideas about nature. I can imagine that someday physicists might tell us that it is time to stop believing in photons, because they have discovered a better way to think about light—but the file will likely live on.

The file is a set of philosophical ideas made into eternal flesh. The ideas expressed by the file include the notion that human expression comes in severable chunks that can be organized as leaves on an abstract tree—and that the chunks have versions and need to be matched to compatible applications.

What do files mean to the future of human expression? This is a harder question to answer than the question "How does the English language influence the thoughts of native English speakers?" At least you can compare English speakers to Chinese speakers, but files are universal. The idea of the file has become so big that we are unable to conceive of a frame large enough to fit around it in order to assess it empirically.

What Happened to Trains, Files, and Musical Notes Could Happen Soon to the Definition of a Human Being

It's worth trying to notice when philosophies are congealing into locked-in software. For instance, is pervasive anonymity or pseudonymity a good thing? It's an important question, because the corresponding philosophies of how humans can express meaning have been so ingrained into the interlocked software designs of the internet that we might never be able to fully get rid of them, or even remember that things could have been different.

We ought to at least try to avoid this particularly tricky example of impending lock-in. Lock-in makes us forget the lost freedoms we had in the digital past. That can make it harder to see the freedoms we have in the digital present. Fortunately, difficult as it is, we can still try to change some expressions of philosophy that are on the verge of becoming locked in place in the tools we use to understand one another and the world.

A Happy Surprise

The rise of the web was a rare instance when we learned new, positive information about human potential. Who would have guessed (at least at first) that millions of people would put so much effort into a project without the presence of advertising, commercial motive, threat of punishment, charismatic figures, identity politics, exploitation of the fear of death, or any of the other classic motivators of mankind. In vast numbers, people did something coopcratively, solely because it was a good idea, and it was beautiful.

Some of the more wild-eyed eccentrics in the digital world had guessed that it would happen—but even so it was a shock when it actually did come to pass. It turns out that even an optimistic, idealistic philosophy is realizable. Put a happy philosophy of life in software, and it might very well come true!

Technology Criticism Shouldn't Be Left to the Luddites

But not all surprises have been happy.

This digital revolutionary still believes in most of the lovely deep ideals that energized our work so many years ago. At the core was a sweet faith in human nature. If we empowered individuals, we believed, more good than harm would result.

The way the internet has gone sour since then is truly perverse. The central faith of the web's early design has been superseded by a different faith in the centrality of imaginary entities epitomized by the idea that the internet as a whole is coming alive and turning into a superhuman creature.

The designs guided by this new, perverse kind of faith put people back in the shadows. The fad for anonymity has undone the great opening-of-everyone's-windows of the 1990s. While that reversal has empowered sadists to a degree, the worst effect is a degradation of ordinary people.

Part of why this happened is that volunteerism proved to be an extremely powerful force in the first iteration of the web. When businesses rushed in to capitalize on what had happened, there was something of a problem, in that the content aspect of the web, the cultural side, was functioning rather well without a business plan.

Google came along with the idea of linking advertising and searching, but that business stayed out of the middle of what people actually did online. It had indirect effects, but not direct ones. The early waves of web activity were remarkably energetic and had a personal quality. People created personal "homepages," and each of them was different, and often strange. The web had flavor.

Entrepreneurs naturally sought to create products that would inspire demand (or at least hypothetical advertising opportunities that might someday compete with Google) where there was no lack to be addressed and no need to be filled, other than greed. Google had discovered a new permanently entrenched niche enabled by the nature of digital technology. It turns out that the digital system of representing people and ads so they can be matched is like MIDI. It is an example of how digital technology can cause an explosive increase in the importance of the "network effect." Every element in the system—every computer, every person, every bit—comes to depend on relentlessly detailed adherence to a common standard, a common point of exchange.

Unlike MIDI, Google's secret software standard is hidden in its computer cloud* instead of being replicated in your pocket. Anyone who wants to place ads must use it, or be out in the cold, relegated to a tiny, irrelevant subculture, just as digital musicians must use MIDI in order to work together in the digital realm. In the case of Google, the monopoly is opaque and proprietary. (Sometimes locked-in digital niches are

---

*"Cloud" is a term for a vast computing resource available over the internet. You never know where the cloud resides physically. Google, Microsoft, IBM, and various government agencies are some of the proprietors of computing clouds.

proprietary, and sometimes they aren't. The dynamics are the same in either case, though the commercial implications can be vastly different.)

There can be only one player occupying Google's persistent niche, so most of the competitive schemes that came along made no money. Behemoths like Facebook have changed the culture with commercial intent, but without, as of this time of writing, commercial achievement.*

In my view, there were a large number of ways that new commercial successes might have been realized, but the faith of the nerds guided entrepreneurs on a particular path. Voluntary productivity had to be commoditized, because the type of faith I'm criticizing thrives when you can pretend that computers do everything and people do nothing.

An endless series of gambits backed by gigantic investments encouraged young people entering the online world for the first time to create standardized presences on sites like Facebook. Commercial interests promoted the widespread adoption of standardized designs like the blog, and these designs encouraged pseudonymity in at least some aspects of their designs, such as the comments, instead of the proud extroversion that characterized the first wave of web culture.

Instead of people being treated as the sources of their own creativity, commercial aggregation and abstraction sites presented anonymized fragments of creativity as products that might have fallen from the sky or been dug up from the ground, obscuring the true sources.

Tribal Accession

The way we got here is that one subculture of technologists has recently become more influential than the others. The winning subculture doesn't have a formal name, but I've sometimes called the members "cybernetic totalists" or "digital Maoists."

---

*Facebook does have advertising, and is surely contemplating a variety of other commercial plays, but so far has earned only a trickle of income, and no profits. The same is true for most of the other web 2.0 businesses. Because of the enhanced network effect of all things digital, it's tough for any new player to become profitable in advertising, since Google has already seized a key digital niche (its ad exchange). In the same way, it would be extraordinarily hard to start a competitor to eBay or Craigslist. Digital network architectures naturally incubate monopolies. That is precisely why the idea of the noosphere, or a collective brain formed by the sum of all the people connected on the internet, has to be resisted with more force than it is promoted.

The ascendant tribe is composed of the folks from the open culture/Creative Commons world, the Linux community, folks associated with the artificial intelligence approach to computer science, the web 2.0 people, the anticontext file sharers and remashers, and a variety of others. Their capital is Silicon Valley, but they have power bases all over the world, wherever digital culture is being created. Their favorite blogs include Boing Boing, TechCrunch, and Slashdot, and their embassy in the old country is *Wired*.

Obviously, I'm painting with a broad brush; not every member of the groups I mentioned subscribes to every belief I'm criticizing. In fact, the groupthink problem I'm worried about isn't so much in the minds of the technologists themselves, but in the minds of the users of the tools the cybernetic totalists are promoting.

The central mistake of recent digital culture is to chop up a network of individuals so finely that you end up with a mush. You then start to care about the abstraction of the network more than the real people who are networked, even though the network by itself is meaningless. Only the people were ever meaningful.

When I refer to the tribe, I am not writing about some distant "them." The members of the tribe are my lifelong friends, my mentors, my students, my colleagues, and my fellow travelers. Many of my friends disagree with me. It is to their credit that I feel free to speak my mind, knowing that I will still be welcome in our world.

On the other hand, I know there is also a distinct tradition of computer science that is humanistic. Some of the better-known figures in this tradition include the late Joseph Weizenbaum, Ted Nelson, Terry Winograd, Alan Kay, Bill Buxton, Doug Englebart, Brian Cantwell Smith, Henry Fuchs, Ken Perlin, Ben Schneiderman (who invented the idea of clicking on a link), and Andy Van Dam, who is a master teacher and has influenced generations of protégés, including Randy Pausch. Another important humanistic computing figure is David Gelernter, who conceived of a huge portion of the technical underpinnings of what has come to be called cloud computing, as well as many of the potential practical applications of clouds.

And yet, it should be pointed out that humanism in computer science doesn't seem to correlate with any particular cultural style. For instance,

Ted Nelson is a creature of the 1960s, the author of what might have been the first rock musical (*Anything & Everything*), something of a vagabond, and a counterculture figure if ever there was one. David Gelernter, on the other hand, is a cultural and political conservative who writes for journals like *Commentary* and teaches at Yale. And yet I find inspiration in the work of them both.

Trap for a Tribe

The intentions of the cybernetic totalist tribe are good. They are simply following a path that was blazed in earlier times by well-meaning Freudians and Marxists—and I don't mean that in a pejorative way. I'm thinking of the earliest incarnations of Marxism, for instance, before Stalinism and Maoism killed millions.

Movements associated with Freud and Marx both claimed foundations in rationality and the scientific understanding of the world. Both perceived themselves to be at war with the weird, manipulative fantasies of religions. And yet both invented their own fantasies that were just as weird.

The same thing is happening again. A self-proclaimed materialist movement that attempts to base itself on science starts to look like a religion rather quickly. It soon presents its own eschatology and its own revelations about what is really going on—portentous events that no one but the initiated can appreciate. The Singularity and the noosphere, the idea that a collective consciousness emerges from all the users on the web, echo Marxist social determinism and Freud's calculus of perversions. We rush ahead of skeptical, scientific inquiry at our peril, just like the Marxists and Freudians.

Premature mystery reducers are rent by schisms, just like Marxists and Freudians always were. They find it incredible that I perceive a commonality in the membership of the tribe. To them, the systems Linux and UNIX are completely different, for instance, while to me they are coincident dots on a vast canvas of possibilities, even if much of the canvas is all but forgotten by now.

At any rate, the future of religion will be determined by the quirks of the software that gets locked in during the coming decades, just like the futures of musical notes and personhood.

Where We Are on the Journey

It's time to take stock. Something amazing happened with the introduction of the World Wide Web. A faith in human goodness was vindicated
when a remarkably open and unstructured information tool was made
available to large numbers of people. That openness can, at this point, be
declared "locked in" to a significant degree. Hurray!

At the same time, some not-so-great ideas about life and meaning were
also locked in, like MIDI's nuance-challenged conception of musical
sound and UNIX's inability to cope with time as humans experience it.

These are acceptable costs, what I would call aesthetic losses. They
are counterbalanced, however, by some aesthetic victories. The digital
world looks better than it sounds because a community of digital
activists, including folks from Xerox Parc (especially Alan Kay), Apple,
Adobe, and the academic world (especially Stanford's Don Knuth)
fought the good fight to save us from the rigidly ugly fonts and other
visual elements we'd have been stuck with otherwise.

Then there are those recently conceived elements of the future of
human experience, like the already locked-in idea of the file, that are as
fundamental as the air we breathe. The file will henceforth be one of the
basic underlying elements of the human story, like genes. We will never
know what that means, or what alternatives might have meant.

On balance, we've done wonderfully well! But the challenge on the
table now is unlike previous ones. The new designs on the verge of being
locked in, the web 2.0 designs, actively demand that people define themselves downward. It's one thing to launch a limited conception of music
or time into the contest for what philosophical idea will be locked in. It is
another to do that with the very idea of what it is to be a person.

Why It Matters

If you feel fine using the tools you use, who am I to tell you that there is
something wrong with what you are doing? But consider these points:

▶ Emphasizing the crowd means deemphasizing individual
  humans in the design of society, and when you ask people
  not to be people, they revert to bad moblike behaviors.

This leads not only to empowered trolls, but to a generally unfriendly and unconstructive online world.

▶ Finance was transformed by computing clouds. Success in finance became increasingly about manipulating the cloud at the expense of sound financial principles.

▶ There are proposals to transform the conduct of science along similar lines. Scientists would then understand less of what they do.

▶ Pop culture has entered into a nostalgic malaise. Online culture is dominated by trivial mashups of the culture that existed before the onset of mashups, and by fandom responding to the dwindling outposts of centralized mass media. It is a culture of reaction without action.

▶ Spirituality is committing suicide. Consciousness is attempting to will itself out of existence.

It might seem as though I'm assembling a catalog of every possible thing that could go wrong with the future of culture as changed by technology, but that is not the case. All of these examples are really just different aspects of one singular, big mistake.

The deep meaning of personhood is being reduced by illusions of bits. Since people will be inexorably connecting to one another through computers from here on out, we must find an alternative.

We have to think about the digital layers we are laying down now in order to benefit future generations. We should be optimistic that civilization will survive this challenging century, and put some effort into creating the best possible world for those who will inherit our efforts.

Next to the many problems the world faces today, debates about online culture may not seem that pressing. We need to address global warming, shift to a new energy cycle, avoid wars of mass destruction, support aging populations, figure out how to benefit from open markets without being disastrously vulnerable to their failures, and take care of other basic business. But digital culture and related topics like the future of privacy and copyrights concern the society we'll have if we can survive these challenges.

20

Every save-the-world cause has a list of suggestions for "what each of us can do": bike to work, recycle, and so on.

I can propose such a list related to the problems I'm talking about:

▸ Don't post anonymously unless you really might be in danger.

▸ If you put effort into Wikipedia articles, put even more effort into using your personal voice and expression outside of the wiki to help attract people who don't yet realize that they are interested in the topics you contributed to.

▸ Create a website that expresses something about who you are that won't fit into the template available to you on a social networking site.

▸ Post a video once in a while that took you one hundred times more time to create than it takes to view.

▸ Write a blog post that took weeks of reflection before you heard the inner voice that needed to come out.

▸ If you are twittering, innovate in order to find a way to describe your internal state instead of trivial external events, to avoid the creeping danger of believing that objectively described events define you, as they would define a machine.

These are some of the things you can do to be a person instead of a source of fragments to be exploited by others.

There are aspects to all these software designs that could be retained more humanistically. A design that shares Twitter's feature of providing ambient continuous contact between people could perhaps drop Twitter's adoration of fragments. We don't really know, because it is an unexplored design space.

As long as you are not defined by software, you are helping to broaden the identity of the ideas that will get locked in for future generations. In most arenas of human expression, it's fine for a person to love the medium they are given to work in. Love paint if you are a painter; love a clarinet if you are a musician. Love the English language (or hate it). Love of these things is a love of mystery.

But in the case of digital creative materials, like MIDI, UNIX, or even the World Wide Web, it's a good idea to be skeptical. These designs came together very recently, and there's a haphazard, accidental quality to them. Resist the easy grooves they guide you into. If you love a medium made of software, there's a danger that you will become entrapped in someone else's recent careless thoughts. Struggle against that!

The Importance of Digital Politics

There was an active campaign in the 1980s and 1990s to promote visual elegance in software. That political movement bore fruit when it influenced engineers at companies like Apple and Microsoft who happened to have a chance to steer the directions software was taking before lock-in made their efforts moot.

That's why we have nice fonts and flexible design options on our screens. It wouldn't have happened otherwise. The seemingly unstoppable mainstream momentum in the world of software engineers was pulling computing in the direction of ugly screens, but that fate was avoided before it was too late.

A similar campaign should be taking place now, influencing engineers, designers, businesspeople, and everyone else to support humanistic alternatives whenever possible. Unfortunately, however, the opposite seems to be happening.

Online culture is filled to the brim with rhetoric about what the true path to a better world ought to be, and these days it's strongly biased toward an antihuman way of thinking.

The Future

The true nature of the internet is one of the most common topics of online discourse. It is remarkable that the internet has grown enough to contain the massive amount of commentary about its own nature.

The promotion of the latest techno-political-cultural orthodoxy, which I am criticizing, has become unceasing and pervasive. The *New York*

Times, for instance, promotes so-called open digital politics on a daily basis even though that ideal and the movement behind it are destroying the newspaper, and all other newspapers.* It seems to be a case of journalistic Stockholm syndrome.

There hasn't yet been an adequate public rendering of an alternative worldview that opposes the new orthodoxy. In order to oppose orthodoxy, I have to provide more than a few jabs. I also have to realize an alternative intellectual environment that is large enough to roam in. Someone who has been immersed in orthodoxy needs to experience a figure-ground reversal in order to gain perspective. This can't come from encountering just a few heterodox thoughts, but only from a new encompassing architecture of interconnected thoughts that can engulf a person with a different worldview.

So, in this book, I have spun a long tale of belief in the opposites of computationalism, the noosphere, the Singularity, web 2.0, the long tail, and all the rest. I hope the volume of my contrarianism will foster an alternative mental environment, where the exciting opportunity to start creating a new digital humanism can begin.

An inevitable side effect of this project of deprogramming through immersion is that I will direct a sustained stream of negativity onto the ideas I am criticizing. Readers, be assured that the negativity eventually tapers off, and that the last few chapters are optimistic in tone.

---

*Today, for instance, as I write these words, there was a headline about R, a piece of geeky statistical software that would never have received notice in the *Times* if it had not been "free." R's nonfree competitor Stata was not even mentioned. (Ashlee Vance, "Data Analysts Captivated by R's Power," *New York Times,* January 6, 2009.)

an apocalypse of self-abdication

THE IDEAS THAT I hope will not be locked in rest on a philosophical foundation that I sometimes call cybernetic totalism. It applies metaphors from certain strains of computer science to people and the rest of reality. Pragmatic objections to this philosophy are presented.

What Do You Do When the Techies Are Crazier Than the Luddites?

The Singularity is an apocalyptic idea originally proposed by John von Neumann, one of the inventors of digital computation, and elucidated by figures such as Vernor Vinge and Ray Kurzweil.

There are many versions of the fantasy of the Singularity. Here's the one Marvin Minsky used to tell over the dinner table in the early 1980s: One day soon, maybe twenty or thirty years into the twenty-first century, computers and robots will be able to construct copies of themselves, and these copies will be a little better than the originals because of intelligent software. The second generation of robots will then make a third, but it will take less time, because of the improvements over the first generation.

The process will repeat. Successive generations will be ever smarter and will appear ever faster. People might think they're in control, until one fine day the rate of robot improvement ramps up so quickly that superintelligent robots will suddenly rule the Earth.

In some versions of the story, the robots are imagined to be microscopic, forming a "gray goo" that eats the Earth; or else the internet itself comes alive and rallies all the net-connected machines into an army to control the affairs of the planet. Humans might then enjoy immortality within virtual reality, because the global brain would be so huge that it would be absolutely easy—a no-brainer, if you will—for it to host all our consciousnesses for eternity.

The coming Singularity is a popular belief in the society of technologists. Singularity books are as common in a computer science department as Rapture images are in an evangelical bookstore.

(Just in case you are not familiar with the Rapture, it is a colorful belief in American evangelical culture about the Christian apocalypse. When I was growing up in rural New Mexico, Rapture paintings would often be found in places like gas stations or hardware stores. They would usually include cars crashing into each other because the virtuous drivers had suddenly disappeared, having been called to heaven just before the onset of hell on Earth. The immensely popular *Left Behind* novels also describe this scenario.)

There might be some truth to the ideas associated with the Singularity at the very largest scale of reality. It might be true that on some vast cosmic basis, higher and higher forms of consciousness inevitably arise, until the whole universe becomes a brain, or something along those lines. Even at much smaller scales of millions or even thousands of years, it is more exciting to imagine humanity evolving into a more wonderful state than we can presently articulate. The only alternatives would be extinction or stodgy stasis, which would be a little disappointing and sad, so let us hope for transcendence of the human condition, as we now understand it.

The difference between sanity and fanaticism is found in how well the believer can avoid confusing consequential differences in timing. If you believe the Rapture is imminent, fixing the problems of this life might not be your greatest priority. You might even be eager to embrace wars and tolerate poverty and disease in others to bring about the conditions that could prod the Rapture into being. In the same way, if you believe the Singularity is coming soon, you might cease to design technology to serve humans, and prepare instead for the grand events it will bring.

But in either case, the rest of us would never know if you had been right. Technology working well to improve the human condition is detectable, and you can see that possibility portrayed in optimistic science fiction like *Star Trek*.

The Singularity, however, would involve people dying in the flesh and being uploaded into a computer and remaining conscious, or people simply being annihilated in an imperceptible instant before a new super-consciousness takes over the Earth. The Rapture and the Singularity share one thing in common: they can never be verified by the living.

You Need Culture to Even Perceive Information Technology

Ever more extreme claims are routinely promoted in the new digital climate. Bits are presented as if they were alive, while humans are transient fragments. Real people must have left all those anonymous comments on blogs and video clips, but who knows where they are now, or if they are dead? The digital hive is growing at the expense of individuality.

Kevin Kelly says that we don't need authors anymore, that all the ideas of the world, all the fragments that used to be assembled into coherent books by identifiable authors, can be combined into one single, global book. *Wired* editor Chris Anderson proposes that science should no longer seek theories that scientists can understand, because the digital cloud will understand them better anyway.[*]

Antihuman rhetoric is fascinating in the same way that self-destruction is fascinating: it offends us, but we cannot look away.

The antihuman approach to computation is one of the most baseless ideas in human history. A computer isn't even there unless a person experiences it. There will be a warm mass of patterned silicon with electricity coursing through it, but the bits don't mean anything without a cultured person to interpret them.

This is not solipsism. You can believe that your mind makes up the world, but a bullet will still kill you. A virtual bullet, however, doesn't

[*]Chris Anderson, "The End of Theory," *Wired*, June 23, 2008 (www.wired.com/science/discoveries/magazine/16-07/pb_theory).

even exist unless there is a person to recognize it as a representation of a bullet. Guns are real in a way that computers are not.

Making People Obsolete So That Computers Seem More Advanced

Many of today's Silicon Valley intellectuals seem to have embraced what used to be speculations as certainties, without the spirit of unbounded curiosity that originally gave rise to them. Ideas that were once tucked away in the obscure world of artificial intelligence labs have gone mainstream in tech culture. The first tenet of this new culture is that all of reality, including humans, is one big information system. That doesn't mean we are condemned to a meaningless existence. Instead there is a new kind of manifest destiny that provides us with a mission to accomplish. The meaning of life, in this view, is making the digital system we call reality function at ever-higher "levels of description."

People pretend to know what "levels of description" means, but I doubt anyone really does. A web page is thought to represent a higher level of description than a single letter, while a brain is a higher level than a web page. An increasingly common extension of this notion is that the net as a whole is or soon will be a higher level than a brain.

There's nothing special about the place of humans in this scheme. Computers will soon get so big and fast and the net so rich with information that people will be obsolete, either left behind like the characters in Rapture novels or subsumed into some cyber-superhuman something.

Silicon Valley culture has taken to enshrining this vague idea and spreading it in the way that only technologists can. Since implementation speaks louder than words, ideas can be spread in the designs of software. If you believe the distinction between the roles of people and computers is starting to dissolve, you might express that—as some friends of mine at Microsoft once did—by designing features for a word processor that are supposed to know what you want, such as when you want to start an outline within your document. You might have had the experience of having Microsoft Word suddenly determine, at the wrong moment, that you are creating an indented outline. While I am all for the automation of petty tasks, this is different.

From my point of view, this type of design feature is nonsense, since you end up having to work more than you would otherwise in order to manipulate the software's expectations of you. The real function of the feature isn't to make life easier for people. Instead, it promotes a new philosophy: that the computer is evolving into a life-form that can understand people better than people can understand themselves.

Another example is what I call the "race to be most meta." If a design like Facebook or Twitter depersonalizes people a little bit, then another service like Friendfeed—which may not even exist by the time this book is published—might soon come along to aggregate the previous layers of aggregation, making individual people even more abstract, and the illusion of high-level metaness more celebrated.

Information Doesn't Deserve to Be Free

"Information wants to be free." So goes the saying. Stewart Brand, the founder of the *Whole Earth Catalog*, seems to have said it first.

I say that information doesn't deserve to be free.

Cybernetic totalists love to think of the stuff as if it were alive and had its own ideas and ambitions. But what if information is inanimate? What if it's even less than inanimate, a mere artifact of human thought? What if only humans are real, and information is not?

Of course, there is a technical use of the term "information" that refers to something entirely real. This is the kind of information that's related to entropy. But that fundamental kind of information, which exists independently of the culture of an observer, is not the same as the kind we can put in computers, the kind that supposedly wants to be free.

Information is alienated experience.

You can think of culturally decodable information as a potential form of experience, very much as you can think of a brick resting on a ledge as storing potential energy. When the brick is prodded to fall, the energy is revealed. That is only possible because it was lifted into place at some point in the past.

In the same way, stored information might cause experience to be revealed if it is prodded in the right way. A file on a hard disk does indeed contain information of the kind that objectively exists. The fact that the bits are discernible instead of being scrambled into mush—the way heat scrambles things—is what makes them bits.

But if the bits can potentially mean something to someone, they can only do so if they are experienced. When that happens, a commonality of culture is enacted between the storer and the retriever of the bits. Experience is the only process that can de-alienate information.

Information of the kind that purportedly wants to be free is nothing but a shadow of our own minds, and wants nothing on its own. It will not suffer if it doesn't get what it wants.

But if you want to make the transition from the old religion, where you hope God will give you an afterlife, to the new religion, where you hope to become immortal by getting uploaded into a computer, then you have to believe information is real and alive. So for you, it will be important to redesign human institutions like art, the economy, and the law to reinforce the perception that information is alive. You demand that the rest of us live in your new conception of a state religion. You need us to deify information to reinforce your faith.

The Apple Falls Again

It's a mistake with a remarkable origin. Alan Turing articulated it, just before his suicide.

Turing's suicide is a touchy subject in computer science circles. There's an aversion to talking about it much, because we don't want our founding father to seem like a tabloid celebrity, and we don't want his memory trivialized by the sensational aspects of his death.

The legacy of Turing the mathematician rises above any possible sensationalism. His contributions were supremely elegant and foundational. He gifted us with wild leaps of invention, including much of the mathematical underpinnings of digital computation. The highest award in computer science, our Nobel Prize, is named in his honor.

Turing the cultural figure must be acknowledged, however. The first thing to understand is that he was one of the great heroes of World War II. He was the first "cracker," a person who uses computers to defeat an enemy's security measures. He applied one of the first computers to break a Nazi secret code, called Enigma, which Nazi mathematicians had believed was unbreakable. Enigma was decoded by the Nazis in the field using a mechanical device about the size of a cigar box. Turing reconceived it as a pattern of bits that could be analyzed in a computer, and cracked it wide open. Who knows what world we would be living in today if Turing had not succeeded?

The second thing to know about Turing is that he was gay at a time when it was illegal to be gay. British authorities, thinking they were doing the most compassionate thing, coerced him into a quack medical treatment that was supposed to correct his homosexuality. It consisted, bizarrely, of massive infusions of female hormones.

In order to understand how someone could have come up with that plan, you have to remember that before computers came along, the steam engine was a preferred metaphor for understanding human nature. All that sexual pressure was building up and causing the machine to malfunction, so the opposite essence, the female kind, ought to balance it out and reduce the pressure. This story should serve as a cautionary tale. The common use of computers, as we understand them today, as sources for models and metaphors of ourselves is probably about as reliable as the use of the steam engine was back then.

Turing developed breasts and other female characteristics and became terribly depressed. He committed suicide by lacing an apple with cyanide in his lab and eating it. Shortly before his death, he presented the world with a spiritual idea, which must be evaluated separately from his technical achievements. This is the famous Turing test. It is extremely rare for a genuinely new spiritual idea to appear, and it is yet another example of Turing's genius that he came up with one.

Turing presented his new offering in the form of a thought experiment, based on a popular Victorian parlor game. A man and a woman hide, and a judge is asked to determine which is which by relying only on the texts of notes passed back and forth.

Turing replaced the woman with a computer. Can the judge tell which is the man? If not, is the computer conscious? Intelligent? Does it deserve equal rights?

It's impossible for us to know what role the torture Turing was enduring at the time played in his formulation of the test. But it is undeniable that one of the key figures in the defeat of fascism was destroyed, by our

WHAT IS A PERSON?

side, after the war, because he was gay. No wonder his imagination pondered the rights of strange creatures.

When Turing died, software was still in such an early state that no one knew what a mess it would inevitably become as it grew. Turing imagined a pristine, crystalline form of existence in the digital realm, and I can imagine it might have been a comfort to imagine a form of life apart from the torments of the body and the politics of sexuality. It's notable that it is the woman who is replaced by the computer, and that Turing's suicide echoes Eve's fall.

The Turing Test Cuts Both Ways

Whatever the motivation, Turing authored the first trope to support the idea that bits can be alive on their own, independent of human observers. This idea has since appeared in a thousand guises, from artificial intelligence to the hive mind, not to mention many overhyped Silicon Valley start-ups.

It seems to me, however, that the Turing test has been poorly interpreted by generations of technologists. It is usually presented to support the idea that machines can attain whatever quality it is that gives people consciousness. After all, if a machine fooled you into believing it was conscious, it would be bigoted for you to still claim it was not.

What the test really tells us, however, even if it's not necessarily what Turing hoped it would say, is that machine intelligence can only be known in a relative sense, in the eyes of a human beholder.*

The AI way of thinking is central to the ideas I'm criticizing in this

---

*One extension of the tragedy of Turing's death is that he didn't live long enough to articulate all that he probably would have about his own point of view on the Turing test.

Historian George Dyson suggests that Turing might have sided *against* the cybernetic totalists. For instance, here is an excerpt from a paper Turing wrote in 1939, titled "Systems of Logic Based on Ordinals": "We have been trying to see how far it is possible to eliminate intuition, and leave only ingenuity. We do not mind how much ingenuity is required, and therefore assume it to be available in unlimited supply." The implication seems to be that we are wrong to imagine that ingenuity can be infinite, even with computing clouds, so therefore intuition will never be made obsolete.

Turing's 1950 paper on the test includes this extraordinary passage: "In attempting to construct such machines we should not be irreverently usurping His power of creating souls, any more than we are in the procreation of children: rather we are, in either case, instruments of His will providing mansions for the souls that He creates."

book. If a machine can be conscious, then the computing cloud is going to be a better and far more capacious consciousness than is found in an individual person. If you believe this, then working for the benefit of the cloud over individual people puts you on the side of the angels.

But the Turing test cuts both ways. You can't tell if a machine has gotten smarter or if you've just lowered your own standards of intelligence to such a degree that the machine seems smart. If you can have a conversation with a simulated person presented by an AI program, can you tell how far you've let your sense of personhood degrade in order to make the illusion work for you?

People degrade themselves in order to make machines seem smart all the time. Before the crash, bankers believed in supposedly intelligent algorithms that could calculate credit risks before making bad loans. We ask teachers to teach to standardized tests so a student will look good to an algorithm. We have repeatedly demonstrated our species' bottomless ability to lower our standards to make information technology look good. Every instance of intelligence in a machine is ambiguous.

The same ambiguity that motivated dubious academic AI projects in the past has been repackaged as mass culture today. Did that search engine really know what you want, or are you playing along, lowering your standards to make it seem clever? While it's to be expected that the human perspective will be changed by encounters with profound new technologies, the exercise of treating machine intelligence as real requires people to reduce their mooring to reality.

A significant number of AI enthusiasts, after a protracted period of failed experiments in tasks like understanding natural language, eventually found consolation in the adoration for the hive mind, which yields better results because there are real people behind the curtain.

Wikipedia, for instance, works on what I call the Oracle illusion, in which knowledge of the human authorship of a text is suppressed in order to give the text superhuman validity. Traditional holy books work in precisely the same way and present many of the same problems.

This is another of the reasons I sometimes think of cybernetic totalist culture as a new religion. The designation is much more than an approximate metaphor, since it includes a new kind of quest for an afterlife. It's so weird to me that Ray Kurzweil wants the global computing cloud to scoop up the contents of our brains so we can live forever in virtual reality. When my friends and I built the first virtual reality machines, the whole point was to make this world more creative, expressive, empathic, and interesting. It was not to escape it.

A parade of supposedly distinct "big ideas" that amount to the worship of the illusions of bits has enthralled Silicon Valley, Wall Street, and other centers of power. It might be Wikipedia or simulated people on the other end of the phone line. But really we are just hearing Turing's mistake repeated over and over.

Or Consider Chess

Will trendy cloud-based economics, science, or cultural processes outpace old-fashioned approaches that demand human understanding? No, because it is only encounters with human understanding that allow the contents of the cloud to exist.

Fragment liberation culture breathlessly awaits future triumphs of technology that will bring about the Singularity or other imaginary events. But there are already a few examples of how the Turing test has been approximately passed, and has reduced personhood. Chess is one.

The game of chess possesses a rare combination of qualities: it is easy to understand the rules, but it is hard to play well; and, most important, the urge to master it seems timeless. Human players achieve ever higher levels of skill, yet no one will claim that the quest is over.

Computers and chess share a common ancestry. Both originated as tools of war. Chess began as a battle simulation, a mental martial art. The design of chess reverberates even further into the past than that—all the way back to our sad animal ancestry of pecking orders and competing clans.

Likewise, modern computers were developed to guide missiles and break secret military codes. Chess and computers are both direct descendants of the violence that drives evolution in the natural world, however sanitized and abstracted they may be in the context of civilization. The drive to compete is palpable in both computer science and chess, and when they are brought together, adrenaline flows.

What makes chess fascinating to computer scientists is precisely that we're bad at it. From our point of view, human brains routinely do things that seem almost insuperably difficult, like understanding sentences—yet we don't hold sentence-comprehension tournaments, because we find that task too easy, too ordinary.

Computers fascinate and frustrate us in a similar way. Children can learn to program them, yet it is extremely difficult for even the most accomplished professional to program them well. Despite the evident potential of computers, we know full well that we have not thought of the best programs to write.

But all of this is not enough to explain the outpouring of public angst on the occasion of Deep Blue's victory in May 1997 over world chess champion Gary Kasparov, just as the web was having its first major influences on popular culture. Regardless of all the old-media hype, it was clear that the public's response was genuine and deeply felt. For millennia, mastery of chess had indicated the highest, most refined intelligence—and now a computer could play better than the very best human.

There was much talk about whether human beings were still special, whether computers were becoming our equal. By now, this sort of thing wouldn't be news, since people have had the AI way of thinking pounded into their heads so much that it is sounding like believable old news. The AI way of framing the event was unfortunate, however. What happened was primarily that a team of computer scientists built a very fast machine and figured out a better way to represent the problem of how to choose the next move in a chess game. People, not machines, performed this accomplishment.

The Deep Blue team's central victory was one of clarity and elegance of thought. In order for a computer to beat the human chess champion, two kinds of progress had to converge: an increase in raw hardware power and an improvement in the sophistication and clarity with which the decisions of chess play are represented in software. This dual path made it hard to predict the year, but not the eventuality, that a computer would triumph.

If the Deep Blue team had not been as good at the software problem, a computer would still have become the world champion at some later date, thanks to sheer brawn. So the suspense lay in wondering not whether a chess-playing computer would ever beat the best human chess player, but to what degree programming elegance would play a role in the victory. Deep Blue won earlier than it might have, scoring a point for elegance.

The public reaction to the defeat of Kasparov left the computer science community with an important question, however. Is it useful to portray computers themselves as intelligent or humanlike in any way? Does this presentation serve to clarify or to obscure the role of computers in our lives?

Whenever a computer is imagined to be intelligent, what is really happening is that humans have abandoned aspects of the subject at hand in order to remove from consideration whatever the computer is blind to. This happened to chess itself in the case of the Deep Blue–Kasparov tournament.

There is an aspect of chess that is a little like poker—the staring down of an opponent, the projection of confidence. Even though it is relatively easier to write a program to "play" poker than to play chess, poker is really a game centering on the subtleties of nonverbal communication between people, such as bluffing, hiding emotion, understanding your opponents' psychologies, and knowing how to bet accordingly. In the wake of Deep Blue's victory, the poker side of chess has been largely overshadowed by the abstract, algorithmic aspect—while, ironically, it was in the poker side of the game that Kasparov failed critically.

Kasparov seems to have allowed himself to be spooked by the computer, even after he had demonstrated an ability to defeat it on occasion. He might very well have won if he had been playing a human player with exactly the same move-choosing skills as Deep Blue (or at least as Deep Blue existed in 1997). Instead, Kasparov detected a sinister stone face where in fact there was absolutely nothing. While the contest was not intended as a Turing test, it ended up as one, and Kasparov was fooled.

As I pointed out earlier, the idea of AI has shifted the psychological projection of adorable qualities from computer programs alone to a different target: computer-plus-crowd constructions. So, in 1999 a wikilike crowd of people, including chess champions, gathered to play Kasparov in an online game called "Kasparov versus the World." In this case Kasparov won, though many believe that it was only because of back-stabbing between members of the crowd. We technologists are ceaselessly intrigued by rituals in which we attempt to pretend that people are obsolete.

The attribution of intelligence to machines, crowds of fragments, or other nerd deities obscures more than it illuminates. When people are told that a computer is intelligent, they become prone to changing themselves in order to make the computer appear to work better, instead of demanding that the computer be changed to become more useful. People already tend to defer to computers, blaming themselves when a digital gadget or online service is hard to use.

Treating computers as intelligent, autonomous entities ends up standing the process of engineering on its head. We can't afford to respect our own designs so much.

The Circle of Empathy

The most important thing to ask about any technology is how it changes people. And in order to ask that question I've used a mental device called the "circle of empathy" for many years. Maybe you'll find it useful as well. (The Princeton philosopher often associated with animal rights, Peter Singer, uses a similar term and idea, seemingly a coincident coinage.)

An imaginary circle of empathy is drawn by each person. It circumscribes the person at some distance, and corresponds to those things in the world that deserve empathy. I like the term "empathy" because it has spiritual overtones. A term like "sympathy" or "allegiance" *might* be more precise, but I want the chosen term to be slightly mystical, to suggest that we might not be able to fully understand what goes on between us and others, that we should leave open the possibility that the relationship can't be represented in a digital database.

If someone falls within your circle of empathy, you wouldn't want to see him or her killed. Something that is clearly outside the circle is fair game. For instance, most people would place all other people within the circle, but most of us are willing to see bacteria killed when we brush our teeth, and certainly don't worry when we see an inanimate rock tossed aside to keep a trail clear.

The tricky part is that some entities reside close to the edge of the circle. The deepest controversies often involve whether something or someone should lie just inside or just outside the circle. For instance, the idea of slavery depends on the placement of the slave outside the circle, to make some people nonhuman. Widening the circle to include all people and end slavery has been one of the epic strands of the human story—and it isn't quite over yet.

A great many other controversies fit well in the model. The fight over abortion asks whether a fetus or embryo should be in the circle or not, and the animal rights debate asks the same about animals.

When you change the contents of your circle, you change your conception of yourself. The center of the circle shifts as its perimeter is changed. The liberal impulse is to expand the circle, while conservatives tend to want to restrain or even contract the circle.

Empathy Inflation and Metaphysical Ambiguity

Are there any legitimate reasons not to expand the circle as much as possible? There are.

To expand the circle indefinitely can lead to oppression, because the rights of potential entities (as perceived by only some people) can conflict with the rights of indisputably real people. An obvious example of this is found in the abortion debate. If outlawing abortions did not involve commandeering control of the bodies of other people (pregnant women, in this case), then there wouldn't be much controversy. We would find an easy accommodation.

Empathy inflation can also lead to the lesser, but still substantial, evils of incompetence, trivialization, dishonesty, and narcissism. You cannot live, for example, without killing bacteria. Wouldn't you be projecting your own fantasies on single-cell organisms that would be indifferent to them at best? Doesn't it really become about you instead of the cause at that point? Do you go around blowing up other people's toothbrushes? Do you think the bacteria you saved are morally equivalent to former slaves—and if you do, haven't you diminished the status of those human beings? Even if you can follow your passion to free and protect the world's bacteria with a pure heart, haven't you divorced yourself from the reality of interdependence and transience of all things? You can try to avoid killing bacteria on special occasions, but you need to kill them to live. And even if you are willing to die for your cause, you can't prevent bacteria from devouring your own body when you die.

Obviously the example of bacteria is extreme, but it shows that the circle is only meaningful if it is finite. If we lose the finitude, we lose our own center and identity. The fable of the Bacteria Liberation Front can serve as a parody of any number of extremist movements on the left or the right.

At the same time, I have to admit that I find it impossible to come to a definitive position on many of the most familiar controversies. I am all for animal rights, for instance, but only as a hypocrite. I eat chicken, but I can't eat cephalopods—octopus and squid—because I admire their neurological evolution so intensely. (Cephalopods also suggest an alternate way to think about the long-term future of technology that avoids certain moral dilemmas—something I'll explain later in the book.)

How do I draw my circle? I just spend time with the various species and decide if they feel like they are in my circle or not. I've raised chickens and somehow haven't felt empathy toward them. They are little more than feathery servo-controlled mechanisms compared to goats, for instance, which I have also raised, and will not eat. On the other hand, a colleague of mine, virtual reality researcher Adrian Cheok, feels such empathy with chickens that he built teleimmersion suits for them so that he could telecuddle them from work. We all have to live with our imperfect ability to discern the proper boundaries of our circles of empathy. There will always be cases where reasonable people will disagree. I don't go around telling other people not to eat cephalopods or goats.

The border between person and nonperson might be found somewhere in the embryonic sequence from conception to baby, or in the development of the young child, or the teenager. Or it might be best defined in the phylogenetic path from ape to early human, or perhaps in the cultural history of ancient peasants leading to modern citizens. It might exist somewhere in a continuum between small and large computers. It might have to do with which thoughts you have; maybe self-reflective thoughts or the moral capacity for empathy makes you human. These are some of the many gates to personhood that have been proposed, but none of them seem definitive to me. The borders of person-hood remain variegated and fuzzy.

Paring the Circle

Just because we are unable to know precisely where the circle of empa-thy should lie does not mean that we are unable to know anything at all about it. If we are only able to be approximately moral, that doesn't mean we should give up trying to be moral at all. The term "morality" is usu-ally used to describe our treatment of others, but in this case I am apply-ing it to ourselves just as much.

The dominant open digital culture places digital information process-ing in the role of the embryo as understood by the religious right, or the bacteria in my reductio ad absurdum fable. The error is classical, but the consequences are new. I fear that we are beginning to design ourselves to suit digital models of us, and I worry about a leaching of empathy and humanity in that process.

The rights of embryos are based on extrapolation, while the rights of a competent adult person are as demonstrable as anything can be, since people speak for themselves. There are plenty of examples where it's hard to decide where to place faith in personhood because a proposed being, while it might be deserving of empathy, cannot speak for itself.

Should animals have the same rights as humans? There are special perils when some people hear voices, and extend empathy, that others do not. If it's at all possible, these are exactly the situations that must be left to people close to a given situation, because otherwise we'll ruin per-sonal freedom by enforcing metaphysical ideas on one another.

In the case of slavery, it turned out that, given a chance, slaves could not just speak for themselves, they could speak intensely and well. Moses was unambiguously a person. Descendants of more recent slaves, like Martin Luther King Jr., demonstrated transcendent eloquence and empathy.

The new twist in Silicon Valley is that some people—very influential people—believe they are hearing algorithms and crowds and other internet-supported nonhuman entities speak for themselves. I don't hear those voices, though—and I believe those who do are fooling themselves.

Thought Experiments:
The Ship of Theseus Meets the
Infinite Library of Borges

To help you learn to doubt the fantasies of the cybernetic totalists, I offer two dueling thought experiments.

The first one has been around a long time. As Daniel Dennett tells it: Imagine a computer program that can simulate a neuron, or even a network of neurons. (Such programs have existed for years and in fact are getting quite good.) Now imagine a tiny wireless device that can send and receive signals to neurons in the brain. Crude devices a little like this already exist; years ago I helped Joe Rosen, a reconstructive plastic surgeon at Dartmouth Medical School, build one—the "nerve chip," which was an early attempt to route around nerve damage using prosthetics.

To get the thought experiment going, hire a neurosurgeon to open your skull. If that's an inconvenience, swallow a nano-robot that can perform neurosurgery. Replace one nerve in your brain with one of those wireless gadgets. (Even if such gadgets were already perfected, connecting them would not be possible today. The artificial neuron would have to engage all the same synapses—around seven thousand, on average— as the biological nerve it replaced.)

Next, the artificial neuron will be connected over a wireless link to a simulation of a neuron in a nearby computer. Every neuron has unique chemical and structural characteristics that must be included in the program. Do the same with your remaining neurons. There are between 100 billion and 200 billion neurons in a human brain, so even at only a second per neuron, this will require tens of thousands of years.

Now for the big question: Are you still conscious after the process has been completed?

Furthermore, because the computer is completely responsible for the dynamics of your brain, you can forgo the physical artificial neurons and let the neuron-control programs connect with one another through software alone. Does the computer then become a person? If you believe in consciousness, is your consciousness now in the computer, or perhaps in the software? The same question can be asked about souls, if you believe in them.

Bigger Borges

Here's a second thought experiment. It addresses the same question from the opposite angle. Instead of changing the program running on the computer, it changes the design of the computer.

First, imagine a marvelous technology: an array of flying laser scanners that can measure the trajectories of all the hailstones in a storm. The scanners send all the trajectory information to your computer via a wireless link.

What would anyone do with this data? As luck would have it, there's a wonderfully geeky store in this thought experiment called the Ultimate Computer Store, which sells a great many designs of computers. In fact, every possible computer design that has fewer than some really large number of logic gates is kept in stock.

You arrive at the Ultimate Computer Store with a program in hand. A salesperson gives you a shopping cart, and you start trying out your program on various computers as you wander the aisles. Once in a while you're lucky, and the program you brought from home will run for a reasonable period of time without crashing on a computer. When that happens, you drop the computer in the shopping cart.

For a program, you could even use the hailstorm data. Recall that a computer program is nothing but a list of numbers; there must be some computers in the Ultimate Computer Store that will run it! The strange thing is that each time you find a computer that runs the hailstorm data as a program, the program does something different.

After a while, you end up with a few million word processors, some amazing video games, and some tax-preparation software—all the same program, as it runs on different computer designs. This takes time; in the real world the universe probably wouldn't support conditions for life long enough for you to make a purchase. But this is a thought experiment, so don't be picky.

The rest is easy. Once your shopping cart is filled with a lot of computers that run the hailstorm data, settle down in the store's café. Set up the computer from the first thought experiment, the one that's running a copy of your brain. Now go through all your computers and compare what each one does with what the computer from the first experiment

41

does. Do this until you find a computer that runs the hailstorm data as a program equivalent to your brain.

How do you know when you've found a match? There are endless options. For mathematical reasons, you can never be absolutely sure of what a big program does or if it will crash, but if you found a way to be satisfied with the software neuron replacements in the first thought experiment, you have already chosen your method to approximately evaluate a big program. Or you could even find a computer in your cart that interprets the motion of the hailstorm over an arbitrary period of time as equivalent to the activity of the brain program over a period of time. That way, the dynamics of the hailstorm are matched to the brain program beyond just one moment in time.

After you've done all this, is the hailstorm now conscious? Does it have a soul?

The Metaphysical Shell Game

The alternative to sprinkling magic dust on people is sprinkling it on computers, the hive mind, the cloud, the algorithm, or some other cybernetic object. The right question to ask is, Which choice is crazier?

If you try to pretend to be certain that there's no mystery in something like consciousness, the mystery that is there can pop out elsewhere in an inconvenient way and ruin your objectivity as a scientist. You enter into a metaphysical shell game that can make you dizzy. For instance, you can propose that consciousness is an illusion, but by definition consciousness is the one thing that isn't reduced if it is an illusion.

There's a way that consciousness and time are bound together. If you try to remove any potential hint of mysteriousness from consciousness, you end up mystifying time in an absurd way.

Consciousness is situated in time, because you can't experience a lack of time, and you can't experience the future. If consciousness isn't anything but a false thought in the computer that is your brain, or the universe, then what exactly *is* it that is situated in time? The present moment, the only other thing that could be situated in time, must in that case be a freestanding object, independent of the way it is experienced.

The present moment is a rough concept, from a scientific point of view, because of relativity and the latency of thoughts moving in the brain. We have no means of defining either a single global physical present moment or a precise cognitive present moment. Nonetheless, there must be *some* anchor, perhaps a very fuzzy one, somewhere, somehow, for it to be possible to even speak of it.

Maybe you could imagine the present moment as a metaphysical marker traveling through a timeless version of reality, in which the past and the future are already frozen in place, like a recording head moving across a hard disk.

If you are certain the experience of time is an illusion, all you have left is time itself. *Something* has to be situated—in a kind of metatime or something—in order for the illusion of the present moment to take place at all. You force yourself to say that time itself travels through reality. This is an absurd, circular thought.

To call consciousness an illusion is to give time a supernatural quality—maybe some kind of spooky nondeterminism. Or you can choose a different shell in the game and say that time is natural (not supernatural), and that the present moment is only a possible concept because of consciousness.

The mysterious stuff can be shuffled around, but it is best to just admit when some trace of mystery remains, in order to be able to speak as clearly as possible about the many things that can actually be studied or engineered methodically.

I acknowledge that there are dangers when you allow for the legitimacy of a metaphysical idea (like the potential for consciousness to be something beyond computation). No matter how careful you are not to "fill in" the mystery with superstitions, you might encourage some fundamentalists or new-age romantics to cling to weird beliefs. "Some dreadlocked computer scientist says consciousness might be more than a computer? Then my food supplement must work!"

But the danger of an engineer pretending to know more than he really does is the greater danger, especially when he can reinforce the illusion through the use of computation. The cybernetic totalists awaiting the Singularity are nuttier than the folks with the food supplements.

The Zombie Army

Do fundamental metaphysical—or supposedly antimetaphysical—beliefs trickle down into the practical aspects of our thinking or our personalities? They do. They can turn a person into what philosophers call a "zombie."

Zombies are familiar characters in philosophical thought experiments. They are like people in every way except that they have no internal experience. They are unconscious, but give no externally measurable evidence of that fact. Zombies have played a distinguished role as fodder in the rhetoric used to discuss the mind-body problem and consciousness research. There has been much debate about whether a true zombie could exist, or if internal subjective experience inevitably colors either outward behavior or measurable events in the brain in some way.

I claim that there is one measurable difference between a zombie and a person: a zombie has a different philosophy. Therefore, zombies can only be detected if they happen to be professional philosophers. A philosopher like Daniel Dennett is obviously a zombie.

Zombies and the rest of us do not have a symmetrical relationship. Unfortunately, it is only possible for nonzombies to observe the telltale sign of zombiehood. To zombies, everyone looks the same.

If there are enough zombies recruited into our world, I worry about the potential for a self-fulfilling prophecy. Maybe if people pretend they are not conscious or do not have free will—or that the cloud of online people is a person; if they pretend there is nothing special about the perspective of the individual—then perhaps we have the power to make it so. We might be able to collectively achieve antimagic.

Humans are free. We can commit suicide for the benefit of a Singularity. We can engineer our genes to better support an imaginary hive mind. We can make culture and journalism into second-rate activities and spend centuries remixing the detritus of the 1960s and other eras from before individual creativity went out of fashion.

Or we can believe in ourselves. By chance, it might turn out we are real.

CHAPTER 3

THe nOOSPHeRe IS JUST anOTHeR name FOR eVeRYOne'S InneR TROLL

SOME OF THE fantasy objects arising from cybernetic totalism (like the noosphere, which is a supposed global brain formed by the sum of all the human brains connected through the internet) happen to motivate infelicitous technological designs. For instance, designs that celebrate the noosphere tend to energize the inner troll, or bad actor, within humans.

The Moral Imperative to Create the Blandest Possible Bible

According to a new creed, we technologists are turning ourselves, the planet, our species, everything, into computer peripherals attached to the great computing clouds. The news is no longer about us but about the big new computational object that is greater than us.

The colleagues I disagree with often conceive our discussions as being a contest between a Luddite (who, me?) and the future. But there is more than one possible technological future, and the debate should be about how to best identify and act on whatever freedoms of choice we still have, not about who's the Luddite.

Some people say that doubters of the one true path, like myself, are like the shriveled medieval church officials who fought against poor Johannes Gutenberg's press. We are accused of fearing change, just as the medieval Church feared the printing press. (We might also be told that we are the sort who would have repressed Galileo or Darwin.)

What these critics forget is that printing presses in themselves provide no guarantee of an enlightened outcome. People, not machines, made the Renaissance. The printing that takes place in North Korea today, for instance, is nothing more than propaganda for a personality cult. What is important about printing presses is not the mechanism, but the authors.

An impenetrable tone deafness rules Silicon Valley when it comes to the idea of authorship. This was as clear as ever when John Updike and Kevin Kelly exchanged words on the question of authorship in 2006. Kevin suggested that it was not just a good thing, but a "moral imperative" that all the world's books would soon become effectively "one book" once they were scanned, searchable, and remixable in the universal computational cloud.

Updike used the metaphor of the edges of the physical paper in a physical book to communicate the importance of enshrining the edges between individual authors. It was no use. Doctrinaire web 2.0 enthusiasts only perceived that Updike was being sentimental about an ancient technology.

The approach to digital culture I abhor would indeed turn all the world's books into one book, just as Kevin suggested. It might start to happen in the next decade or so. Google and other companies are scanning library books into the cloud in a massive Manhattan Project of cultural digitization. What happens next is what's important. If the books in the cloud are accessed via user interfaces that encourage mashups of fragments that obscure the context and authorship of each fragment, there will be only one book. This is what happens today with a lot of content; often you don't know where a quoted fragment from a news story came from, who wrote a comment, or who shot a video. A continuation of the present trend will make us like various medieval religious empires, or like North Korea, a society with a single book.*

---

*The Bible can serve as a prototypical example. Like Wikipedia, the Bible's authorship was shared, largely anonymous, and cumulative, and the obscurity of the individual authors served to create an oracle-like ambience for the document as "the literal word of God." If we take a nonmetaphysical view of the Bible, it serves as a link to our ancestors, a window

The ethereal, digital replacement technology for the printing press happens to have come of age in a time when the unfortunate ideology I'm criticizing dominates technological culture. Authorship—the very idea of the individual point of view—is not a priority of the new ideology.

The digital flattening of expression into a global mush is not presently enforced from the top down, as it is in the case of a North Korean printing press. Instead, the design of software builds the ideology into those actions that are the easiest to perform on the software designs that are becoming ubiquitous. It is true that by using these tools, individuals can author books or blogs or whatever, but people are encouraged by the economics of free content, crowd dynamics, and lord aggregators to serve up fragments instead of considered whole expressions or arguments. The efforts of authors are appreciated in a manner that erases the boundaries between them.

The one collective book will absolutely not be the same thing as the library of books by individuals it is bankrupting. Some believe it will be better; others, including me, believe it will be disastrously worse. As the famous line goes from *Inherit the Wind:* "The Bible is a book . . . but it is not the only book." Any singular, exclusive book, even the collective one accumulating in the cloud, will become a cruel book if it is the only one available.

Nerd Reductionism

One of the first printed books that wasn't a bible was 1499's *Hypnerotomachia Poliphili,* or "Poliphili's Strife of Love in a Dream," an illustrated, erotic, occult adventure through fantastic architectural settings. What is most interesting about this book, which looks and reads like a virtual reality fantasy, is that something fundamental about its approach to life—its intelligence, its worldview—is alien to the Church and the Bible.

---

into human nature and our cultural origins, and can be used as a source of solace and inspiration. Someone who believes in a personal God can felicitously believe that the Bible reflects that God indirectly, through the people who wrote it. But when people buy into the oracle illusion, the Bible just turns into a tool to help religious leaders and politicians manipulate them.

It's easy to imagine an alternate history in which everything that was printed on early presses went through the Church and was conceived as an extension of the Bible. "Strife of Love" might have existed in this alternate world, and might have been quite similar. But the "slight" modifications would have consisted of trimming the alien bits. The book would no longer have been as strange. And that tiny shift, even if it had been minuscule in terms of word count, would have been tragic.

This is what happened when elements of indigenous cultures were preserved but de-alienated by missionaries. We know a little about what Aztec or Inca music sounded like, for instance, but the bits that were trimmed to make the music fit into the European idea of church song were the most precious bits. The alien bits are where the flavor is found. They are the portals to strange philosophies. What a loss to not know how New World music would have sounded alien to us! Some melodies and rhythms survived, but the whole is lost.

Something like missionary reductionism has happened to the internet with the rise of web 2.0. The strangeness is being leached away by the mush-making process. Individual web pages as they first appeared in the early 1990s had the flavor of personhood. MySpace preserved some of that flavor, though a process of regularized formatting had begun. Facebook went further, organizing people into multiple-choice identities, while Wikipedia seeks to erase point of view entirely.

If a church or government were doing these things, it would feel authoritarian, but when technologists are the culprits, we seem hip, fresh, and inventive. People will accept ideas presented in technological form that would be abhorrent in any other form. It is utterly strange to hear my many old friends in the world of digital culture claim to be the true sons of the Renaissance without realizing that using computers to reduce individual expression is a primitive, retrograde activity, no matter how sophisticated your tools are.

Rejection of the Idea of Quality Results in a Loss of Quality

The fragments of human effort that have flooded the internet are perceived by some to form a hive mind, or noosphere. These are some of the terms used to describe what is thought to be a new superintelligence

WHAT IS A PERSON?

that is emerging on a global basis on the net. Some people, like Larry Page, one of the Google founders, expect the internet to come alive at some point, while others, like science historian George Dyson, think that might already have happened. Popular derivative terms like "blogosphere" have become commonplace.

A fashionable idea in technical circles is that quantity not only turns into quality at some extreme of scale, but also does so according to principles we already understand. Some of my colleagues think a million, or perhaps a billion, fragmentary insults will eventually yield wisdom that surpasses that of any well-thought-out essay, so long as sophisticated secret statistical algorithms recombine the fragments. I disagree. A trope from the early days of computer science comes to mind: garbage in, garbage out.

There are so many examples of disdain for the idea of quality within the culture of web 2.0 enthusiasts that it's hard to choose an example. I'll choose hive enthusiast Clay Shirky's idea that there is a vast cognitive surplus waiting to be harnessed.

Certainly there is broad agreement that there are huge numbers of people who are undereducated. Of those who are well educated, many are underemployed. If we want to talk about unmet human potential, we might also mention the huge number of people who are desperately poor. The waste of human potential is overwhelming. But these are not the problems that Shirky is talking about.

What he means is that quantity can overwhelm quality in human expression. Here's a quote, from a speech Shirky gave in April 2008:

> And this is the other thing about the size of the cognitive surplus we're talking about. It's so large that even a small change could have huge ramifications. Let's say that everything stays 99 percent the same, that people watch 99 percent as much television as they used to, but 1 percent of that is carved out for producing and for sharing. The Internet-connected population watches roughly a trillion hours of TV a year . . . One percent of that is 98 Wikipedia projects per year worth of participation.

So how many seconds of salvaged erstwhile television time would need to be harnessed to replicate the achievements of, say, Albert

Einstein? It seems to me that even if we could network all the potential aliens in the galaxy—quadrillions of them, perhaps—and get each of them to contribute some seconds to a physics wiki, we would not replicate the achievements of even one mediocre physicist, much less a great one.

Absent Intellectual Modesty

There are at least two ways to believe in the idea of quality. You can believe there's something ineffable going on within the human mind, or you can believe we just don't understand what quality in a mind is yet, even though we might someday. Either of those opinions allows one to distinguish quantity and quality. In order to confuse quantity and quality, you have to reject both possibilities.

The mere possibility of there being something ineffable about personhood is what drives many technologists to reject the notion of quality. They want to live in an airtight reality that resembles an idealized computer program, in which everything is understood and there are no fundamental mysteries. They recoil from even the hint of a potential zone of mystery or an unresolved seam in one's worldview.

This desire for absolute order usually leads to tears in human affairs, so there is a historical reason to distrust it. Materialist extremists have long seemed determined to win a race with religious fanatics: Who can do the most damage to the most people?

At any rate, there is no evidence that quantity becomes quality in matters of human expression or achievement. What matters instead, I believe, is a sense of focus, a mind in effective concentration, and an adventurous individual imagination that is distinct from the crowd.

Of course, I can't describe what it is that a mind does, because no one can. We don't understand how brains work. We understand a lot about how parts of brains work, but there are fundamental questions that have not even been fully articulated yet, much less answered.

For instance, how does reason work? How does meaning work? The usual ideas currently in play are variations on the notion that pseudo-Darwinian selection goes on within the brain. The brain tries out different thought patterns, and the ones that work best are reinforced. That's awfully vague. But there's no reason that Darwinian evolution could not have given rise to processes within the human brain that jumped out of the Darwinian progression. While the physical brain is a product of evolution as we are coming to understand it, the cultural brain might be a way of transforming the evolved brain according to principles that cannot be explained in evolutionary terms.

Another way to put this is that there might be some form of creativity other than selection. I certainly don't know, but it seems pointless to insist that what we already understand must suffice to explain what we don't understand.

What I'm struck by is the lack of intellectual modesty in the computer science community. We are happy to enshrine into engineering designs mere hypotheses—and vague ones at that—about the hardest and most profound questions faced by science, as if we already possess perfect knowledge.

If it eventually turns out that there is something about an individual human mind that is different from what can be achieved by a noosphere, that "special element" might potentially turn out to have any number of qualities. It is possible that we will have to await scientific advances that will only come in fifty, five hundred, or five thousand years before we can sufficiently appreciate our own brains.

Or it might turn out that a distinction will forever be based on principles we cannot manipulate. This might involve types of computation that are unique to the physical brain, maybe relying on forms of causation that depend on remarkable and nonreplicable physical conditions. Or it might involve software that could only be created by the long-term work of evolution, which cannot be reverse-engineered or mucked with in any accessible way. Or it might even involve the prospect, dreaded by some, of dualism, a reality for consciousness as apart from mechanism.

The point is that we don't know. I love speculating about the workings of the brain. Later in the book, I'll present some thoughts on how to use computational metaphors to at least vaguely imagine how a process like meaning might work in the brain. But I would abhor anyone using my speculations as the basis of a design for a tool to be used by real people. An aeronautical engineer would never put passengers in a plane based on an untested, speculative theory, but computer scientists commit analogous sins all the time.

An underlying problem is that technical people overreact to religious extremists. If a computer scientist says that we don't understand how the brain works, will that empower an ideologue to then claim that some particular religion has been endorsed? This is a real danger, but over-claiming by technical people is the greater danger, since we end up confusing ourselves.

It Is Still Possible to Get Rid of Crowd Ideology in Online Designs

From an engineering point of view, the difference between a social networking site and the web as it existed before such sites were introduced is a matter of small detail. You could always create a list of links to your friends on your website, and you could always send e-mails to a circle of friends announcing whatever you cared to. All that the social networking services offer is a prod to use the web in a particular way, according to a particular philosophy.

If anyone wanted to reconsider social network designs, it would be easy enough to take a standoffish approach to describing what goes on between people. It could be left to people to communicate what they want to say about their relationships in their own way.

If someone wants to use words like "single" or "looking" in a self-description, no one is going to prevent that. Search engines will easily find instances of those words. There's no need for an imposed, official category.

If you read something written by someone who used the term "single" in a custom-composed, unique sentence, you will inevitably get a first whiff of the subtle experience of the author, something you would not get from a multiple-choice database. Yes, it would be a tiny bit more work for everyone, but the benefits of semiautomated self-presentation are illusory. If you start out by being fake, you'll eventually have to put in twice the effort to undo the illusion if anything good is to come of it.

This is an example of a simple way in which digital designers could choose to be modest about their claims to understand the nature of human beings. Enlightened designers leave open the possibility of either metaphysical specialness in humans or in the potential for unforeseen creative processes that aren't explained by ideas like evolution that we already believe we can capture in software systems. That kind of modesty is the signature quality of being human-centered.

There would be trade-offs. Adopting a metaphysically modest approach would make it harder to use database techniques to create instant lists of people who are, say, emo, single, and affluent. But I don't think that would be such a great loss. A stream of misleading information is no asset.

It depends on how you define yourself. An individual who is receiving a flow of reports about the romantic status of a group of friends must learn to think in the terms of the flow if it is to be perceived as worth reading at all. So here is another example of how people are able to lessen themselves so as to make a computer seem accurate. Am I accusing all those hundreds of millions of users of social networking sites of reducing themselves in order to be able to use the services? Well, yes, I am.

I know quite a few people, mostly young adults but not all, who are proud to say that they have accumulated thousands of friends on Facebook. Obviously, this statement can only be true if the idea of friendship is reduced. A real friendship ought to introduce each person to unexpected weirdness in the other. Each acquaintance is an alien, a well of unexplored difference in the experience of life that cannot be imagined or accessed in any way but through genuine interaction. The idea of friendship in database-filtered social networks is certainly reduced from that.

It is also important to notice the similarity between the lords and peasants of the cloud. A hedge fund manager might make money by using the computational power of the cloud to calculate fantastical financial instruments that make bets on derivatives in such a way as to invent out of thin air the phony virtual collateral for stupendous risks. This is a subtle form of counterfeiting, and is precisely the same maneuver a socially competitive teenager makes in accumulating fantastical numbers of "friends" on a service like Facebook.

Ritually Faked Relationships Beckon to Messiahs Who May Never Arrive

But let's suppose you disagree that the idea of friendship is being reduced, and are confident that we can keep straight the two uses of the word, the old use and the new use. Even then one must remember that the customers of social networks are not the members of those networks.

The real customer is the advertiser of the future, but this creature has yet to appear in any significant way as this is being written. The whole artifice, the whole idea of fake friendship, is just bait laid by the lords of the clouds to lure hypothetical advertisers—we might call them messianic advertisers—who could someday show up.

The hope of a thousand Silicon Valley start-ups is that firms like Facebook are capturing extremely valuable information called the "social graph." Using this information, an advertiser might hypothetically be able to target all the members of a peer group just as they are forming their opinions about brands, habits, and so on.

Peer pressure is the great power behind adolescent behavior, goes the reasoning, and adolescent choices become life choices. So if someone could crack the mystery of how to make perfect ads using the social graph, an advertiser would be able to design peer pressure biases in a population of real people who would then be primed to buy whatever the advertiser is selling for their whole lives.

The situation with social networks is layered with multiple absurdities. The advertising idea hasn't made any money so far, because ad dollars appear to be better spent on searches and in web pages. If the revenue never appears, then a weird imposition of a database-as-reality ideology will have colored generations of teen peer group and romantic experiences for no business or other purpose.

If, on the other hand, the revenue does appear, evidence suggests that its impact will be truly negative. When Facebook has attempted to turn the social graph into a profit center in the past, it has created ethical disasters.

A famous example was 2007's Beacon. This was a suddenly imposed feature that was hard to opt out of. When a Facebook user made a purchase anywhere on the internet, the event was broadcast to all the so-called friends in that person's network. The motivation was to find a way to package peer pressure as a service that could be sold to advertisers. But it meant that, for example, there was no longer a way to buy a surprise birthday present. The commercial lives of Facebook users were no longer their own.

The idea was instantly disastrous, and inspired a revolt. The MoveOn network, for instance, which is usually involved in electoral politics, activated its huge membership to complain loudly. Facebook made a quick retreat.

The Beacon episode cheered me, and strengthened my sense that people are still able to steer the evolution of the net. It was one good piece of evidence against metahuman technological determinism. The net doesn't design itself. We design it.

But even after the Beacon debacle, the rush to pour money into social networking sites continued without letup. The only hope for social networking sites from a business point of view is for a magic formula to appear in which some method of violating privacy and dignity becomes acceptable. The Beacon episode proved that this cannot happen too quickly, so the question now is whether the empire of Facebook users can be lulled into accepting it gradually.

The Truth About Crowds

The term "wisdom of crowds" is the title of a book by James Surowiecki and is often introduced with the story of an ox in a marketplace. In the story, a bunch of people all guess the animal's weight, and the average of the guesses turns out to be generally more reliable than any one person's estimate.

A common idea about why this works is that the mistakes various people make cancel one another out; an additional, more important idea is that there's at least a little bit of correctness in the logic and assumptions underlying many of the guesses, so they center around the right answer. (This latter formulation emphasizes that individual intelligence is still at the core of the collective phenomenon.) At any rate, the effect is repeatable and is widely held to be one of the foundations of both market economies and democracies.

People have tried to use computing clouds to tap into this collective wisdom effect with fanatic fervor in recent years. There are, for instance, well-funded—and prematurely well-trusted—schemes to apply stock market–like systems to programs in which people bet on the viability of answers to seemingly unanswerable questions, such as when terrorist events will occur or when stem cell therapy will allow a person to grow new teeth. There is also an enormous amount of energy being put into aggregating the judgments of internet users to create "content," as in the collectively generated link website Digg.

How to Use a Crowd Well

The reason the collective can be valuable is precisely that its peaks of intelligence and stupidity are not the same as the ones usually displayed by individuals.

What makes a market work, for instance, is the marriage of collective and individual intelligence. A marketplace can't exist only on the basis of having prices determined by competition. It also needs entrepreneurs to come up with the products that are competing in the first place.

In other words, clever individuals, the heroes of the marketplace, ask the questions that are answered by collective behavior. They bring the ox to the market.

There are certain types of answers that ought not be provided by an individual. When a government bureaucrat sets a price, for instance, the result is often inferior to the answer that would come from a reasonably informed collective that is reasonably free of manipulation or runaway internal resonances. But when a collective designs a product, you get design by committee, which is a derogatory expression for a reason.

Collectives can be just as stupid as any individual—and, in important cases, stupider. The interesting question is whether it's possible to map out where the one is smarter than the many.

There is a substantial history to this topic, and varied disciplines have accumulated instructive results. Every authentic example of collective intelligence that I am aware of also shows how that collective was guided or inspired by well-meaning individuals. These people focused the collective and in some cases also corrected for some of the common hive mind failure modes. The balancing of influence between people and collectives is the heart of the design of democracies, scientific communities, and many other long-standing success stories.

The preinternet world provides some great examples of how individual human-driven quality control can improve collective intelligence. For example, an independent press provides tasty news about politicians by journalists with strong voices and reputations, like the Watergate reporting of Bob Woodward and Carl Bernstein. Without an independent press, composed of heroic voices, the collective becomes stupid and unreliable, as has been demonstrated in many historical instances—most recently, as many have suggested, during the administration of George W. Bush.

Scientific communities likewise achieve quality through a cooperative process that includes checks and balances, and ultimately rests on a foundation of goodwill and "blind" elitism (blind in the sense that ideally anyone can gain entry, but only on the basis of a meritocracy). The tenure system and many other aspects of the academy are designed to support the idea that individual scholars matter, not just the process or the collective.

Yes, there have been plenty of scandals in government, the academy, and the press. No mechanism is perfect. But still here we are, having benefited from all of these institutions. There certainly have been plenty of bad reporters, self-deluded academic scientists, incompetent bureaucrats, and so on. Can the hive mind help keep them in check? The answer provided by experiments in the preinternet world is yes—but only if some signal processing has been placed in the loop.

Signal processing is a bag of tricks engineers use to tweak flows of information. A common example is the way you can set the treble and bass on an audio signal. If you turn down the treble, you are reducing the amount of energy going into higher frequencies, which are composed of tighter, smaller sound waves. Similarly, if you turn up the bass, you are heightening the biggest, broadest waves of sound.

Some of the regulating mechanisms for collectives that have been most successful in the preinternet world can be understood as being like treble and bass controls. For instance, what if a collective moves too readily and quickly, jittering instead of settling down to provide a stable answer? This happens on the most active Wikipedia entries, for example, and has also been seen in some speculation frenzies in open markets.

One service performed by representative democracy is low-pass filtering, which is like turning up the bass and turning down the treble. Imagine the jittery shifts that would take place if a wiki were put in charge of ·

writing laws. It's a terrifying thing to consider. Superenergized people would be struggling to shift the wording of the tax code on a frantic, never-ending basis. The internet would be swamped.

Such chaos can be avoided in the same way it already is, albeit imperfectly: by the slower processes of elections and court proceedings. These are like bass waves. The calming effect of orderly democracy achieves more than just the smoothing out of peripatetic struggles for consensus. It also reduces the potential for the collective to suddenly jump into an overexcited state when too many rapid changes coincide in such a way that they don't cancel one another out.

For instance, stock markets might adopt automatic trading shutoffs, which are triggered by overly abrupt shifts in price or trading volume. (In Chapter 6 I will tell how Silicon Valley ideologues recently played a role in convincing Wall Street that it could do without some of these checks on the crowd, with disastrous consequences.)

Wikipedia had to slap a crude low-pass filter on the jitteriest entries, such as "President George W. Bush." There's now a limit to how often a particular person can remove someone else's text fragments. I suspect that these kinds of adjustments will eventually evolve into an approximate mirror of democracy as it was before the internet arrived.

The reverse problem can also appear. The hive mind can be on the right track, but moving too slowly. Sometimes collectives can yield brilliant results given enough time—but sometimes there isn't enough time. A problem like global warming might automatically be addressed eventually if the market had enough time to respond to it. (Insurance rates, for instance, would climb.) Alas, in this case there doesn't appear to be enough time, because the market conversation is slowed down by the legacy effect of existing investments. Therefore some other process has to intervene, such as politics invoked by individuals.

Another example of the slow hive problem: there was a lot of technology developed—but very slowly—in the millennia before there was a clear idea of how to be empirical, before we knew how to have a peer-reviewed technical literature and an education based on it, and before there was an efficient market to determine the value of inventions.

What is crucial about modernity is that structure and constraints were part of what sped up the process of technological development, not just pure openness and concessions to the collective. This is an idea that will be examined in Chapter 10.

An Odd Lack of Curiosity

The "wisdom of crowds" effect should be thought of as a tool. The value of a tool is its usefulness in accomplishing a task. The point should never be the glorification of the tool. Unfortunately, simplistic free market ideologues and noospherians tend to reinforce one another's unjustified sentimentalities about their chosen tools.

Since the internet makes crowds more accessible, it would be beneficial to have a wide-ranging, clear set of rules explaining when the wisdom of crowds is likely to produce meaningful results. Surowiecki proposes four principles in his book, framed from the perspective of the interior dynamics of the crowd. He suggests there should be limits on the ability of members of the crowd to see how others are about to decide on a question, in order to preserve independence and avoid mob behavior. Among other safeguards, I would add that a crowd should never be allowed to frame its own questions, and its answers should never be more complicated than a single number or multiple choice answer.

More recently, Nassim Nicholas Taleb has argued that applications of statistics, such as crowd wisdom schemes, should be divided into four quadrants. He defines the dangerous "Fourth Quadrant" as comprising problems that have both complex outcomes and unknown distributions of outcomes. He suggests making that quadrant taboo for crowds.

Maybe if you combined all our approaches you'd get a practical set of rules for avoiding crowd failures. Then again, maybe we are all on the wrong track. The problem is that there's been inadequate focus on the testing of such ideas.

There's an odd lack of curiosity about the limits of crowd wisdom. This is an indication of the faith-based motivations behind such schemes. Numerous projects have looked at how to improve specific markets and other crowd wisdom systems, but too few projects have framed the question in more general terms or tested general hypotheses about how crowd systems work.

Trolls

"Troll" is a term for an anonymous person who is abusive in an online environment. It would be nice to believe that there is a only a minute troll population living among us. But in fact, a great many people have experienced being drawn into nasty exchanges online. Everyone who has experienced that has been introduced to his or her inner troll.

I have tried to learn to be aware of the troll within myself. I notice that I can suddenly become relieved when someone else in an online exchange is getting pounded or humiliated, because that means I'm safe for the moment. If someone else's video is being ridiculed on YouTube, then mine is temporarily protected. But that also means I'm complicit in a mob dynamic. Have I ever planted a seed of mob-beckoning ridicule in order to guide the mob to a target other than myself? Yes, I have, though I shouldn't have. I observe others doing that very thing routinely in anonymous online meeting places.

I've also found that I can be drawn into ridiculous pissing matches online in ways that just wouldn't happen otherwise, and I've never noticed any benefit. There is never a lesson learned, or a catharsis of victory or defeat. If you win anonymously, no one knows, and if you lose, you just change your pseudonym and start over, without having modified your point of view one bit.

If the troll is anonymous and the target is known, then the dynamic is even worse than an encounter between anonymous fragmentary pseudo-people. That's when the hive turns against personhood. For instance, in 2007 a series of "Scarlet Letter" postings in China incited online throngs to hunt down accused adulterers. In 2008, the focus shifted to Tibet sympathizers. Korea has one of the most intense online cultures in the world, so it has also suffered some of the most extreme trolling. Korean movie star Choi Jin-sil, sometimes described as the "Nation's Actress," committed suicide in 2008 after being hounded online by trolls, but she was only the most famous of a series of similar suicides.

In the United States, anonymous internet users have ganged up on targets like Lori Drew, the woman who created a fake boy persona on the internet in order to break the heart of a classmate of her daughter's, which caused the girl to commit suicide.

But more often the targets are chosen randomly, following the pattern described in the short story "The Lottery" by Shirley Jackson. In the story, residents of a placid small town draw lots to decide which individual will be stoned to death each year. It is as if a measure of human cruelty must be released, and to do so in a contained yet random way limits the damage by using the fairest possible method.

Some of the better-known random victims of troll mobs include the blogger Kathy Sierra. She was suddenly targeted in a multitude of ways, such as having images of her as a sexually mutilated corpse posted prominently, apparently in the hopes that her children would see them. There was no discernible reason Sierra was targeted. Her number was somehow drawn from the lot.

Another famous example is the tormenting of the parents of Mitchell Henderson, a boy who committed suicide. They were subjected to gruesome audio-video creations and other tools at the disposal of virtual sadists. Another occurence is the targeting of epileptic people with flashing web designs in the hope of inducing seizures.

There is a vast online flood of videos of humiliating assaults on helpless victims. The culture of sadism online has its own vocabulary and has gone mainstream. The common term "lulz," for instance, refers to the gratification of watching others suffer over the cloud.*

When I criticize this type of online culture, I am often accused of being either an old fart or an advocate of censorship. Neither is the case. I don't think I'm necessarily any better, or more moral, than the people who tend the lulzy websites. What I'm saying, though, is that the user interface designs that arise from the ideology of the computing cloud make people—all of us—less kind. Trolling is not a string of isolated incidents, but the status quo in the online world.

---

*A website called the Encyclopedia Dramatica brags on its main page that it "won the 2nd Annual Mashable Open Web Awards for the wiki category." As I check it today, in late 2008, just as this book is about to leave my hands, the headlining "Article of the Now" is described in this way: "[Three guys] decided that the best way to commemorate their departing childhood was to kill around 21 people with hammers, pipes and screwdrivers, and record the whole thing on their [video recording] phones." This story was also featured on Boing Boing—which went to the trouble of determining that it was not a hoax—and other top sites this week.

The Standard Sequence
of Troll Invocation

There are recognizable stages in the degradation of anonymous, fragmentary communication. If no pack has emerged, then individuals start to fight. This is what happens all the time in online settings. A later stage appears once a pecking order is established. Then the members of the pack become sweet and supportive of one another, even as they goad one another into ever more intense hatred of nonmembers.

This suggests a hypothesis to join the ranks of ideas about how the circumstances of our evolution influenced our nature. We, the big-brained species, probably didn't get that way to fill a single, highly specific niche. Instead, we must have evolved with the ability to switch between different niches. We evolved to be *both* loners *and* pack members. We are optimized not so much to be one or the other, but to be able to switch between them.

New patterns of social connection that are unique to online culture have played a role in the spread of modern networked terrorism. If you look at an online chat about anything, from guitars to poodles to aerobics, you'll see a consistent pattern: jihadi chat looks just like poodle chat. A pack emerges, and either you are with it or against it. If you join the pack, then you join the collective ritual hatred.

If we are to continue to focus the powers of digital technology on the project of making human affairs less personal and more collective, then we ought to consider how that project might interact with human nature.

The genetic aspects of behavior that have received the most attention (under rubrics like sociobiology or evolutionary psychology) have tended to focus on things like gender differences and mating behaviors, but my guess is that clan orientation and its relationship to violence will turn out to be the most important area of study.

Design Underlies Ethics
in the Digital World

People are not universally nasty online. Behavior varies considerably from site to site. There are reasonable theories about what brings out the best or worst online behaviors: demographics, economics, child-rearing trends, perhaps even the average time of day of usage could play a role. My opinion, however, is that certain details in the design of the user interface experience of a website are the most important factors.

People who can spontaneously invent a pseudonym in order to post a comment on a blog or on YouTube are often remarkably mean. Buyers and sellers on eBay are a little more civil, despite occasional disappointments, such as encounters with flakiness and fraud. Based on those data, you could conclude that it isn't exactly anonymity, but *transient* anonymity, coupled with a lack of consequences, that brings out online idiocy.

With more data, that hypothesis can be refined. Participants in Second Life (a virtual online world) are generally not quite as mean to one another as are people posting comments to Slashdot (a popular technology news site) or engaging in edit wars on Wikipedia, even though all allow pseudonyms. The difference might be that on Second Life the pseudonymous personality itself is highly valuable and requires a lot of work to create.

So a better portrait of the troll-evoking design is effortless, consequence-free, transient anonymity in the service of a goal, such as promoting a point of view, that stands entirely apart from one's identity or personality. Call it drive-by anonymity.

Computers have an unfortunate tendency to present us with binary choices at every level, not just at the lowest one, where the bits are switching. It is easy to be anonymous or fully revealed, but hard to be revealed just enough. Still, that does happen, to varying degrees. Sites like eBay and Second Life give hints about how design can promote a middle path.

Anonymity certainly has a place, but that place needs to be designed carefully. Voting and peer review are preinternet examples of beneficial anonymity. Sometimes it is desirable for people to be free of fear of reprisal or stigma in order to invoke honest opinions. To have a substantial exchange, however, you need to be fully present. That is why facing one's accuser is a fundamental right of the accused.

Could Drive-by Anonymity Scale Up the Way Communism and Fascism Did?

For the most part, the net has delivered happy surprises about human potential. As I pointed out earlier, the rise of the web in the early 1990s took place without leaders, ideology, advertising, commerce, or anything other than a positive sensibility shared by millions of people. Who would have thought that was possible? Ever since, there has been a constant barrage of utopian extrapolations from positive online events. Whenever a blogger humiliates a corporation by posting documentation of an infelicitous service representative, we can expect triumphant hollers about the end of the era of corporate abuses.

It stands to reason, however, that the net can also accentuate negative patterns of behavior or even bring about unforeseen social pathology. Over the last century, new media technologies have often become prominent as components of massive outbreaks of organized violence.

For example, the Nazi regime was a major pioneer of radio and cinematic propaganda. The Soviets were also obsessed with propaganda technologies. Stalin even nurtured a "Manhattan Project" to develop a 3-D theater with incredible, massive optical elements that would deliver perfected propaganda. It would have been virtual reality's evil twin if it had been completed. Many people in the Muslim world have only gained access to satellite TV and the internet in the last decade. These media certainly have contributed to the current wave of violent radicalism. In all these cases, there was an intent to propagandize, but intent isn't everything.

It's not crazy to worry that, with millions of people connected through a medium that sometimes brings out their worst tendencies, massive, fascist-style mobs could rise up suddenly. I worry about the next generation of young people around the world growing up with internet-based technology that emphasizes crowd aggregation, as is the current fad. Will they be more likely to succumb to pack dynamics when they come of age?

What's to prevent the acrimony from scaling up? Unfortunately, history tells us that collectivist ideals can mushroom into large-scale social disasters. The *fascias* and communes of the past started out with small numbers of idealistic revolutionaries.

I am afraid we might be setting ourselves up for a reprise. The recipe that led to social catastrophe in the past was economic humiliation combined with collectivist ideology. We already have the ideology in its new digital packaging, and it's entirely possible we could face dangerously traumatic economic shocks in the coming decades.

An Ideology of Violation

The internet has come to be saturated with an ideology of violation. For instance, when some of the more charismatic figures in the online world, including Jimmy Wales, one of the founders of Wikipedia, and Tim O'Reilly, the coiner of the term "web 2.0," proposed a voluntary code of conduct in the wake of the bullying of Kathy Sierra, there was a widespread outcry, and the proposals went nowhere.

The ideology of violation does not radiate from the lowest depths of trolldom, but from the highest heights of academia. There are respectable academic conferences devoted to methods of violating sanctities of all kinds. The only criterion is that researchers come up with some way of using digital technology to harm innocent people who thought they were safe.

In 2008, researchers from the University of Massachusetts at Amherst and the University of Washington presented papers at two of these conferences (called Defcon and Black Hat), disclosing a bizarre form of attack that had apparently not been expressed in public before, even in works of fiction. They had spent two years of team effort figuring out how to use mobile phone technology to hack into a pacemaker and turn it off by remote control, in order to kill a person. (While they withheld some of the details in their public presentation, they certainly described enough to assure protégés that success was possible.)

The reason I call this an expression of ideology is that there is a strenuously constructed lattice of arguments that decorate this murderous behavior so that it looks grand and new. If the same researchers had done something similar without digital technology, they would at the very least have lost their jobs. Suppose they had spent a couple of years and significant funds figuring out how to rig a washing machine to poison clothing in order to (hypothetically) kill a child once dressed. Or what if they had devoted a lab in an elite university to finding a new way to imperceptibly tamper with skis to cause fatal accidents on the slopes? These are certainly doable projects, but because they are not digital, they don't support an illusion of ethics.

A summary of the ideology goes like this: All those nontechnical, ignorant, innocent people out there are going about their lives thinking that they are safe, when in actuality they are terribly vulnerable to those who are smarter than they are. Therefore, we smartest technical people ought to invent ways to attack the innocents, and publicize our results, so that everyone is alerted to the dangers of our superior powers. After all, a clever evil person might come along.

There are some cases in which the ideology of violation does lead to practical, positive outcomes. For instance, any bright young technical person has the potential to discover a new way to infect a personal computer with a virus. When that happens, there are several possible next steps. The least ethical would be for the "hacker" to infect computers. The most ethical would be for the hacker to quietly let the companies that support the computers know, so that users can download fixes. An intermediate option would be to publicize the "exploit" for glory. A fix can usually be distributed before the exploit does harm.

But the example of the pacemakers is entirely different. The rules of the cloud apply poorly to reality. It took two top academic labs two years of focused effort to demonstrate the exploit, and that was only possible because a third lab at a medical school was able to procure pacemakers and information about them that would normally be very hard to come by. Would high school students or terrorists, or any other imaginable party, have been able to assemble the resources necessary to figure out whether it was possible to kill people in this new way?

The fix in this case would require many surgeries—more than one for each person who wears a pacemaker. New designs of pacemakers will only inspire new exploits. There will always be a new exploit, because there is no such thing as perfect security. Will each heart patient have to schedule heart surgeries on an annual basis in order to keep ahead of academic do-gooders, just in order to stay alive? How much would it cost? How many would die from the side effects of surgery? Given the endless opportunity for harm, no one will be able to act on the information the researchers have graciously provided, so everyone with a pacemaker will forever be at greater risk than they otherwise would have been. No improvement has taken place, only harm.

Those who disagree with the ideology of violation are said to subscribe to a fallacious idea known as "security through obscurity." Smart people aren't supposed to accept this strategy for security, because the internet is supposed to have made obscurity obsolete.

Therefore, another group of elite researchers spent years figuring out how to pick one of the toughest-to-pick door locks, and posted the results on the internet. This was a lock that thieves had not learned to pick on their own. The researchers compared their triumph to Turing's cracking of Enigma. The method used to defeat the lock would have remained obscure were it not for the ideology that has entranced much of the academic world, especially computer science departments.

Surely obscurity is the only fundamental form of security that exists, and the internet by itself doesn't make it obsolete. One way to deprogram academics who buy into the pervasive ideology of violation is to point out that security through obscurity has another name in the world of biology: biodiversity.

The reason some people are immune to a virus like AIDS is that their particular bodies are obscure to the virus. The reason that computer viruses infect PCs more than Macs is not that a Mac is any better engineered, but that it is relatively obscure. PCs are more commonplace. This means that there is more return on the effort to crack PCs.

There is no such thing as an unbreakable lock. In fact, the vast majority of security systems are not too hard to break. But there is always effort required to figure out how to break them. In the case of pacemakers, it took two years at two labs, which must have entailed a significant expense.

Another predictable element of the ideology of violation is that anyone who complains about the rituals of the elite violators will be accused of spreading FUD—fear, uncertainty, and doubt. But actually it's the ideologues who seek publicity. The whole point of publicizing exploits like the attack on pacemakers is the glory. If that notoriety isn't based on spreading FUD, what is?

The MIDI of Anonymity

Just as the idea of a musical note was formalized and rigidified by MIDI, the idea of drive-by, trollish, pack-switch anonymity is being plucked from the platonic realm and made into immovable eternal architecture by software. Fortunately, the process isn't complete yet, so there is still time to promote alternative designs that resonate with human kindness.

When people don't become aware of, or fail to take responsibility for, their role, accidents of time and place can determine the outcomes of the standards wars between digital ideologies. Whenever we notice an instance when history was swayed by accident, we also notice the latitude we have to shape the future.

Hive mind ideology wasn't running the show during earlier eras of the internet's development. The ideology became dominant *after* certain patterns were set, because it sat comfortably with those patterns. The origins of today's outbreaks of nasty online behavior go back quite a way, to the history of the counterculture in America, and in particular to the war on drugs.

Before the World Wide Web, there were other types of online connections, of which Usenet was probably the most influential. Usenet was an online directory of topics where anyone could post comments, drive-by style. One portion of Usenet, called "alt," was reserved for nonacademic topics, including those that were oddball, pornographic, illegal, or offensive. A lot of the alt material was wonderful, such as information about obscure musical instruments, while some of it was sickening, such as tutorials on cannibalism.

To get online in those days you usually had to have an academic, corporate, or military connection, so the Usenet population was mostly adult and educated. That didn't help. Some users still turned into mean idiots online. This is one piece of evidence that it's the design, not the demographic, that concentrates bad behavior. Since there were so few people online, though, bad "netiquette" was then more of a curiosity than a problem.

Why did Usenet support drive-by anonymity? You could argue that it was the easiest design to implement at the time, but I'm not sure that's true. All those academic, corporate, and military users belonged to large, well-structured organizations, so the hooks were immediately available to create a nonanonymous design. If that had happened, today's websites might not have inherited the drive-by design aesthetic.

So if it wasn't laziness that promoted online anonymity, what was it?

Facebook Is Similar to No Child Left Behind

Personal reductionism has always been present in information systems. You have to declare your status in reductive ways when you file a tax return. Your real life is represented by a silly, phony set of database entries in order for you to make use of a service in an approximate way. Most people are aware of the difference between reality and database entries when they file taxes.

But the order is reversed when you perform the same kind of self-reduction in order to create a profile on a social networking site. You fill in the data: profession, marital status, and residence. But in this case digital reduction becomes a causal element, mediating contact between new friends. That is new. It used to be that government was famous for being impersonal, but in a postpersonal world, that will no longer be a distinction.

It might at first seem that the experience of youth is now sharply divided between the old world of school and parents, and the new world of social networking on the internet, but actually school now belongs on the new side of the ledger. Education has gone through a parallel transformation, and for similar reasons.

Information systems need to have information in order to run, but information underrepresents reality. Demand more from information than it can give, and you end up with monstrous designs. Under the No Child Left Behind Act of 2002, for example, U.S. teachers are forced to choose between teaching general knowledge and "teaching to the test." The best teachers are thus often disenfranchised by the improper use of educational information systems.

What computerized analysis of all the country's school tests has done to education is exactly what Facebook has done to friendships. In both cases, life is turned into a database. Both degradations are based on the same philosophical mistake, which is the belief that computers can presently represent human thought or human relationships. These are things computers cannot currently do.

Whether one expects computers to improve in the future is a different issue. In a less idealistic atmosphere it would go without saying that software should only be designed to perform tasks that can be successfully performed at a given time. That is not the atmosphere in which internet software is designed, however.

If we build a computer model of an automobile engine, we know how to test whether it's any good. It turns out to be easy to build bad models! But it is possible to build good ones. We must model the materials, the fluid dynamics, the electrical subsystem. In each case, we have extremely solid physics to rely on, but we have lots of room for making mistakes in the logic or conception of how the pieces fit together. It is inevitably a long, unpredictable grind to debug a serious simulation of any complicated system. I've worked on varied simulations of such things as surgical procedures, and it is a humbling process. A good surgical simulation can take years to refine.

When it comes to people, we technologists must use a completely different methodology. We don't understand the brain well enough to comprehend phenomena like education or friendship on a scientific basis. So when we deploy a computer model of something like learning or friendship in a way that has an effect on real lives, we are relying on faith. When we ask people to live their lives through our models, we are potentially reducing life itself. How can we ever know what we might be losing?

The Abstract Person Obscures
the Real Person

What happened to musical notes with the arrival of MIDI is happening to people.

It breaks my heart when I talk to energized young people who idolize the icons of the new digital ideology, like Facebook, Twitter, Wikipedia, and free/open/Creative Commons mashups. I am always struck by the endless stress they put themselves through. They must manage their online reputations constantly, avoiding the ever-roaming evil eye of the hive mind, which can turn on an individual at any moment. A "Facebook generation" young person who suddenly becomes humiliated online has no way out, for there is only one hive.

I would prefer not to judge the experiences or motivations of other people, but surely this new strain of gadget fetishism is driven more by fear than by love.

At their best, the new Facebook/Twitter enthusiasts remind me of the anarchists and other nutty idealists who populated youth culture when I grew up. The ideas might be silly, but at least the believers have fun as they rebel against the parental-authority quality of entities like record companies that attempt to fight music piracy.

The most effective young Facebook users, however—the ones who will probably be winners if Facebook turns out to be a model of the future they will inhabit as adults—are the ones who create successful online fictions about themselves.

They tend their doppelgängers fastidiously. They must manage offhand remarks and track candid snapshots at parties as carefully as a politician. Insincerity is rewarded, while sincerity creates a lifelong taint. Certainly, some version of this principle existed in the lives of teenagers before the web came along, but not with such unyielding, clinical precision.

The frenetic energy of the original flowering of the web has reappeared in a new generation, but there is a new brittleness to the types of connections people make online. This is a side effect of the illusion that digital representations can capture much about actual human relationships.

The binary character at the core of software engineering tends to reappear at higher levels. It is far easier to tell a program to run or not to run, for instance, than it is to tell it to sort-of run. In the same way, it is easier to set up a rigid representation of human relationships on digital networks: on a typical social networking site, either you are designated to be in a couple or you are single (or you are in one of a few other predetermined states of being)—and that reduction of life is what gets broadcast between friends all the time. What is communicated between people eventually becomes their truth. Relationships take on the troubles of software engineering.

Just a Reminder That I'm Not Anti-Net

It seems ridiculous to have to say this, but just in case anyone is getting the wrong idea, let me affirm that I am not turning against the internet. I love the internet.

For just one example among many, I have been spending quite a lot of time on an online forum populated by oud players. (The oud is a Middle Eastern string instrument.) I hesitate to mention it, because I worry that any special little place on the internet can be ruined if it gets too much attention.

The oud forum revives the magic of the early years of the internet. There's a bit of a feeling of paradise about it. You can feel each participant's passion for the instrument, and we help one another become more intense. It's amazing to watch oud players from around the world cheer on an oud builder as he posts pictures of an instrument under construction. It's thrilling to hear clips from a young player captured in midair just as she is getting good.

The fancy web 2.0 designs of the early twenty-first century start off by classifying people into bubbles, so you meet your own kind. Facebook tops up dating pools, LinkedIn corrals careerists, and so on.

The oud forum does the opposite. There you find Turks and Armenians, elders and kids, Israelis and Palestinians, rich professionals and struggling artists, formal academics and bohemian street musicians, all talking with one another about a shared obsession. We get to know one another; we are not fragments to one another. Inner trolls most definitely appear now and then, but less often than in most online environments. The oud forum doesn't solve the world's problems, but it does allow us to live larger than them.

When I told Kevin Kelly about this magical confluence of obsessive people, he immediately asked if there was a particular magical person who tended the oud forum. The places that work online always turn out to be the beloved projects of individuals, not the automated aggregations of the cloud. In this case, of course, there is such a magical person, who turns out to be a young Egyptian American oud player in Los Angeles.

The engineer in me occasionally ponders the rather crude software that the forum runs on. The deep design mystery of how to organize and present multiple threads of conversation on a screen remains as unsolved as ever. But just when I am about to dive into a design project to improve forum software, I stop and wonder if there really is much room for improvement.

It's the people who make the forum, not the software. Without the software, the experience would not exist at all, so I celebrate that software, as flawed as it is. But it's not as if the forum would really get much better if the software improved. Focusing too much on the software might even make things worse by shifting the focus from the people.

There is huge room for improvement in digital technologies overall. I would love to have telepresence sessions with distant oudists, for instance. But once you have the basics of a given technological leap in place, it's always important to step back and focus on the people for a while.

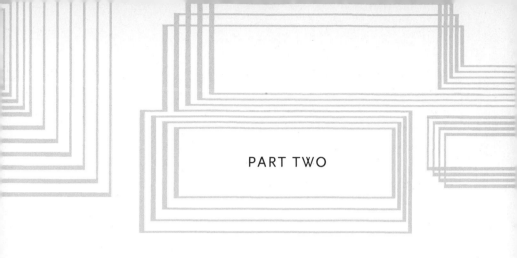

PART TWO

WHAT WILL MONEY BE?

THUS FaR, I have presented two ways in which the current dominant ideology of the digital world, cybernetic totalism, has been a failure.

The first example might be called a spiritual failure. The ideology has encouraged narrow philosophies that deny the mystery of the existence of experience. A practical problem that can trickle down from this mistake is that we become vulnerable to redirecting the leap of faith we call "hope" away from people and toward gadgets.

The second failure is behavioral. It naturally happens that the designs that celebrate the noosphere and other ideals of cybernetic totalism tend to undervalue humans. Examples are the ubiquitous invocations of anonymity and crowd identity. It shouldn't be much of a surprise that these designs tend to reinforce indifferent or poor treatment of humans. In this section, a third failure is presented, this time in the sphere of economics.

For millions of people, the internet means endless free copies of music, videos, and other forms of detached human expression. For a few brilliant and lucky people, the internet has meant an ability to spin financial schemes that were too complex to exist in the past, creating dangerous, temporary illusions of risk-free ways to create money out of thin air.

I will argue that there are similarities and hidden links between these two trends. In each case, there are obvious short-term benefits for some people, but ultimately a disaster for everyone in the long term.

I'll discuss "free culture" first. The disaster related to free culture is still in its early stages. Low-bandwidth forms of human expression, like music and newspaper-style reporting, are already being demoted into a sorry state. High-bandwidth expressions, like movies, are on their way to meeting the same fate.

CHAPTER 4

DIGITAL PEASANT CHIC

ANOTHER PROBLEM WITH the philosophy I am criticizing is that it leads to economic ideas that disfavor the loftiest human avocations. In this and the following sections I will address an orthodoxy that has recently arisen in the world of digital culture and entrepreneurship. Problems associated with overly abstract, complex, and dangerous financial schemes are connected with the ideals of "open" or "free" culture.

Ruining an Appointment with Destiny

The ideology that has overtaken much of the cloud-computing scene— exemplified by causes like free or open culture—has the potential to ruin a moment that has been anticipated since at least as far back as the nineteenth century. Once technological advances are sufficient to potentially offer all people lives filled with health and ease, what will happen? Will only a tiny minority benefit?

While the relative number of desperately poor people is decreasing, income differences between the rich and the poor are increasing at an accelerating rate. The middle zone between wealth and poverty is being stretched, and new seams are likely to appear.

Medicine is on the verge of mastering some of the fundamental mechanisms of aging. Drastic differences in people's wealth will translate into unprecedented, drastic differences in life expectancy. The developed world might start to know how the most abject, hungry, and ill people in the poorest parts of the world feel today. Middle-class life expectancies could start to seem puny compared to those of a lucky elite.

What would happen if you discovered one morning that while a few of your acquaintances who had made or inherited a lot of money had undergone procedures that would extend their life spans by decades, those procedures were too expensive for you and your family? That's the kind of morning that could turn almost anyone into a Marxist.

Marx was all about technological change. Unfortunately, his approach to correcting inequities spawned an awful series of violent revolutions. He argued that the playing field should be leveled before the technologies of abundance mature. It has been repeatedly confirmed, however, that leveling a playing field with a Marxist revolution kills, dulls, or corrupts most of the people on the field. Even so, versions of his ideas continue to have enormous appeal for many, especially young people. Marx's ideas still color utopian technological thinking, including many of the thoughts that appear to be libertarian on the surface. (I will examine stealth technomarxism later on.)

What has saved us from Marxism is simply that new technologies have in general created new jobs—and those jobs have generally been better than the old ones. They have been ever more elevated—more cerebral, creative, cultural, or strategic—than the jobs they replaced. A descendant of a Luddite who smashed looms might be programming robotic looms today.

Crashing Down Maslow's Pyramid

Abraham Maslow was a twentieth-century psychologist who proposed that human beings seek to sate ever more exalted needs as their baser needs are met. A starving person might choose to seek food before social status, for instance, but once a person isn't hungry, a desire for status can become as intense as the earlier quest for food.

Maslow's hierarchy is rooted in the ground, in agriculture and subsistence, but it reaches upward to lofty heights. Sometimes it is visualized as a pyramid, with the base representing the basic needs of survival, like food. The next layer up represents safety, then love/belonging, then esteem, and, finally, as the pyramidion, self-actualization. Self-actualization includes creativity.

Historical improvements in the economic status of ordinary people can be correlated with a climb up Maslow's pyramid. One consequence of ascending the ramp of technological progress, as happened rapidly during industrialization, was that large numbers of people started to make a living from meeting needs at ever higher elevations on Maslow's hierarchy. A vast middle class of teachers, accountants, and, yes, reporters and musicians arose where there had been only a few servants of the royal courts and churches before.

The early generations of Marxists didn't hate these elevated strivers, though they did seek to flatten status in society. Mao brought a different sensibility into play, in which only toil within the foundation layer of Maslow's hierarchy was worthy of reward. The peasants, working in the fields much as they had for millennia, were to be celebrated, while high-altitude creatures, such as intellectuals, were to be punished.

The open culture movement has, weirdly, promoted a revival of this sensibility. Classical Maoism didn't really reject hierarchy; it only suppressed any hierarchy that didn't happen to be the power structure of the ruling Communist Party. In China today, that hierarchy has been blended with others, including celebrity, academic achievement, and personal wealth and status, and China is certainly stronger because of that change.

In the same way, digital Maoism doesn't reject all hierarchy. Instead, it overwhelmingly rewards the one preferred hierarchy of digital metaness, in which a mashup is more important than the sources who were mashed. A blog of blogs is more exalted than a mere blog. If you have seized a very high niche in the aggregation of human expression—in the way that Google has with search, for instance—then you can become superpowerful. The same is true for the operator of a hedge fund. "Meta" equals power in the cloud.

The hierarchy of metaness is the natural hierarchy for cloud gadgets in the same way that Maslow's idea describes a natural hierarchy of human aspirations.

To be fair, open culture is distinct from Maoism in another way. Maoism is usually associated with authoritarian control of the communication of ideas. Open culture is not, although the web 2.0 designs, like wikis, tend to promote the false idea that there is only one universal truth in some arenas where that isn't so.

But in terms of economics, digital Maoism is becoming a more apt term with each passing year. In the physical world, libertarianism and

Maoism are about as different as economic philosophies could be, but in the world of bits, as understood by the ideology of cybernetic totalism, they blur, and are becoming harder and harder to distinguish from each other.

Morality Needs Technology
If It's to Do Any Good

Prior to industrialization, every civilization relied on large classes of people who were slaves or near-slaves. Without technological progress, all the well-meaning political and moral progress in the world wasn't enough to change the conditions of the lives of ordinary people.

Slaves powered even the precocious democracy of ancient Athens. It was only the development of functioning machines, which seemed to amplify mere thoughts into physical actualities, that made slavery obsolete.

I'll go further than that. People will focus on activities other than fighting and killing one another only so long as technologists continue to come up with ways to improve living standards for everyone at once. That isn't to say that technological progress guarantees moral progress. However, expanding wealth is necessary if morality is to have any large-scale effect on events, and improving technology is the only way to expand wealth for many people at the same time.

This hasn't always been as true as it is today. Colonialism and conquest were ways to generate wealth that were distinguishable from technological improvement, though the military and technological domains have always been tightly correlated. The discovery of fresh natural resources, like a new oil field, can also expand wealth. But we can no longer count on forms of wealth expansion outside of technological innovation. The low-hanging fruit have been plucked. Only extreme inventiveness can expand wealth now.

Technological Change Is Stressful

Machines allowed large numbers of people to rise from slave status to skilled-worker status. Nonetheless, one persistent dark side of industrialization is that any skill, no matter how difficult to acquire, can become obsolete when the machines improve.

In the nineteenth century, workers started to wonder what would happen when machines became good enough to function autonomously. Would capitalism have to be retired in order to grant sustenance to the masses of people who were no longer needed to run the machines? Could a fundamental economic transformation of that kind happen peacefully?

So far, each new wave of technological change has brought with it new kinds of demands for human labor. The automobile sent buggy-whip manufacturers into oblivion but employed armies of mechanics. The transformations of labor continue: a sizable number of the employed people in the world are currently tending the untidy bits of the world's computers one way or another. They work at help desks, for enterprise support companies, and in IT departments.

But we are already approaching the endgame for at least some aspects of the coexistence of people and machines. Robots are starting to get better. The semiautonomous rovers on Mars have outperformed all expectations, cute little Roombas are sweeping our floors, and you can buy a car that parks itself.

Robots are even more impressive in the lab. They perform combat missions and surgery and, ominously, fabricate products from raw materials. There are already affordable homemade hobbyist models of small fabricating robots that can create household items on demand right in your house, based on plans downloaded from the net.

The Devaluation of Everything

One of our essential hopes in the early days of the digital revolution was that a connected world would create more opportunities for personal advancement for everyone. Maybe it will eventually, but there has been more of an inverted effect so far, at least in the United States. During the past decade and a half, since the debut of the web, even during the *best* years of the economic boom times, the middle class in the United States declined. Wealth was ever more concentrated.

I'm not saying this is the fault of the net, but if we digital technologists are supposed to be providing a cure, we aren't doing it fast enough.

If we can't reformulate digital ideals before our appointment with destiny, we will have failed to bring about a better world. Instead we will usher in a dark age in which everything human is devalued.

This kind of devaluation will go into high gear when information systems become able to act without constant human intervention in the physical world, through robots and other automatic gadgets. In a crowdsourced world, the peasants of the noosphere will ride a dismal boomerang between gradual impoverishment under robot-driven capitalism and a dangerously sudden, desperate socialism.

The Only Product That Will Maintain Its Value After the Revolution

There is, unfortunately, only one product that can maintain its value as everything else is devalued under the banner of the noosphere. At the end of the rainbow of open culture lies an eternal spring of advertisements. Advertising is elevated by open culture from its previous role as an accelerant and placed at the center of the human universe.

There was a discernible ambient disgust with advertising in an earlier, more hippielike phase of Silicon Valley, before the outlandish rise of Google. Advertising was often maligned back then as a core sin of the bad old-media world we were overthrowing. Ads were at the very heart of the worst of the devils we would destroy, commercial television.

Ironically, advertising is now singled out as the only form of expression meriting genuine commercial protection in the new world to come. Any other form of expression is to be remashed, anonymized, and decontextualized to the point of meaninglessness. Ads, however, are to be made ever more contextual, and the content of the ad is absolutely sacrosanct. No one—and I mean no one—dares to mash up ads served in the margins of their website by Google. When Google started to rise, a common conversation in Silicon Valley would go like this: "Wait, don't we hate advertising?" "Well, we hate *old* advertising. The new kind of advertising is unobtrusive and useful."

The centrality of advertising to the new digital hive economy is absurd, and it is even more absurd that this isn't more generally recognized. The most tiresome claim of the reigning official digital philosophy is that crowds working for free do a better job at some things than paid antediluvian experts. Wikipedia is often given as an example. If that is so—and as I explained, if the conditions are right it sometimes can be—why doesn't the principle dissolve the persistence of advertising as a business?

A functioning, honest crowd-wisdom system ought to trump paid persuasion. If the crowd is so wise, it should be directing each person optimally in choices related to home finance, the whitening of yellow teeth, and the search for a lover. All that paid persuasion ought to be mooted. Every penny Google earns suggests a failure of the crowd—and Google is earning a lot of pennies.

Accelerating a Vacuum

If you want to know what's really going on in a society or ideology, follow the money. If money is flowing to advertising instead of musicians, journalists, and artists, then a society is more concerned with manipulation than truth or beauty. If content is worthless, then people will start to become empty-headed and contentless.

The combination of hive mind and advertising has resulted in a new kind of social contract. The basic idea of this contract is that authors, journalists, musicians, and artists are encouraged to treat the fruits of their intellects and imaginations as fragments to be given without pay to the hive mind. Reciprocity takes the form of self-promotion. Culture is to become precisely nothing but advertising.

It's true that today the idea can work in some situations. There are a few widely celebrated, but exceptional, success stories that have taken on mythical qualities. These stories are only possible because we are in a transitional period, in which a few lucky people can benefit from the best of the old- and new-media worlds at the same time, and the fact of their unlikely origins can be spun into a still-novel marketing narrative.

Thus someone as unlikely as Diablo Cody, who worked as a stripper, can blog and receive enough attention to get a book contract, and then have the opportunity to have her script made into a movie—in this case, the widely acclaimed *Juno*. To think about technologies, however, you have to learn to think as if you're already living in the future.

It is my hope that book publishing will continue remuneratively into the digital realm. But that will only happen if digital designs evolve to make it possible. As things stand, books will be vastly devalued as soon as large numbers of people start reading from an electronic device.

The same is true for movies. Right now, there are still plenty of people in the habit of buying movies on disk, and of going out to movie theaters. This is the way culture works these days. You have to deliver it through some kind of proprietary hardware, like a theater or a paper book, in order to charge for it.

This is not a sustainable solution. The younger you are, the more likely you are to grab a movie for free over the net instead of buying a disk. As for theaters, I wish them a long, healthy continued life, but imagine a world in which a superb fifty-dollar projector can be set up anywhere, in the woods or at the beach, and generate as good an experience. That is the world we will live in within a decade. Once file sharing shrinks Hollywood as it is now shrinking the music companies, the option of selling a script for enough money to make a living will be gone.

Blaming Our Victims

In the early days of so-called open culture, I was an early adopter of one of our talking points that has since become a cliché: All the dinosaurs of the old order have been given fair notice of the digital revolution to come. If they can't adapt, it is due to their own stubbornness, rigidity, or stupidity. Blame them for their fate.

This is what we have said since about our initial victims, like the record companies and newspapers. But none of us was ever able to give the dinosaurs any constructive advice about how to survive. And we miss them now more than we have been willing to admit.

Actually, as long as we put the blame on them, it is okay to admit that we miss the declining "mainstream media." A popular 2008 blog post by Jon Talton blamed newspapers for their own decline, in keeping with the established practices of the revolution. It ended with this stereotypical accusation, which I'll quote at length:

> The biggest problem . . . was the collapse of an unsustainable business model. Simply put, the model involved sending miniskirted saleswomen out to sell ads at confiscatory rates to lecherous old car dealers and appliance-store owners . . .

> *Now the tailspin continues, and the damage to our democracy*
> *is hard to overstate. It's no coincidence that the United States*
> *stumbled into Iraq and is paralyzed before serious challenges at*
> *home and abroad at precisely the moment when real journalism*
> *is besieged. It almost might make the conspiracy minded think*
> *there was a grand plan to keep us dumb.*

Of course, I've selected just one little blog post out of millions. But it is highly representative of the tenor of online commentary. No one's ever been able to offer good advice for the dying newspapers, but it is still considered appropriate to blame them for their own fate.

An important question has been raised by this rant, and it would be taboo to ask it in online circles if it weren't gift wrapped in blanket attacks on the dignity of our victims: Would the recent years of American history have been any different, any less disastrous, if the economic model of the newspaper had not been under assault? We had more bloggers, sure, but also fewer Woodwards and Bernsteins during a period in which ruinous economic and military decisions were made. The Bush years are almost universally perceived as having been catastrophic: the weapons of mass destruction illusion, the economic implosion. Instead of facing up to a tough press, the administration was made vaguely aware of mobs of noisily opposed bloggers nullifying one another. Sure, bloggers uncovered the occasional scandal, but so did opposing bloggers. The effect of the blogosphere overall was a wash, as is always the case for the type of flat open systems celebrated these days.

Peasants and Lords of the Clouds

If some free video of a silly stunt will draw as many eyeballs as the product of a professional filmmaker on a given day, then why pay the filmmaker? If an algorithm can use cloud-based data to unite those eyeballs with the video clip of the moment, why pay editors or impresarios? In the new scheme there is nothing but location, location, location. Rule the computing cloud that routes the thoughts of the hive mind, and you'll be infinitely wealthy!

We already see the effect of an emerging winner-take-all social contract in students. The brightest computer science students are increasingly turning away from intellectually profound aspects of the field and instead hoping to land a spot in the new royalty at the center of the cloud, perhaps programming a hedge fund. Or the best students might be hatching plans to launch a social networking site for affluent golfers. One Ivy League engineering school unofficially banned that idea as a model business plan in a class on entrepreneurship because it had become so commonplace. Meanwhile, creative people—the new peasants—come to resemble animals converging on shrinking oases of old media in a depleted desert.

One effect of the so-called free way of thinking is that it could eventually force anyone who wants to survive on the basis of mental activity (other than cloud tending) to enter into some sort of legal or political fortress—or become a pet of a wealthy patron—in order to be protected from the rapacious hive mind. What free really means is that artists, musicians, writers, and filmmakers will have to cloak themselves within stodgy institutions.

We forget what a wonder, what a breath of fresh air it has been to have creative people make their way in the world of commerce instead of patronage. Patrons gave us Bach and Michelangelo, but it's unlikely patrons would have given us Vladimir Nabokov, the Beatles, or Stanley Kubrick.

CHAPTER 5

THE CITY IS BUILT TO MUSIC

THE FATES OF musicians in the emerging digital economy are examined.

How Long Is Too Long to Wait?

A little over a decade and a half ago, with the birth of the World Wide Web, a clock started. The old-media empires were put on a path of predictable obsolescence. But would a superior replacement arise in time? What we idealists said then was, "Just wait! More opportunities will be created than destroyed." Isn't fifteen years long enough to wait before we switch from hope to empiricism? The time has come to ask, "Are we building the digital utopia for people or machines?" If it's for people, we have a problem.

Open culture revels in bizarre, exaggerated perceptions of the evils of the record companies or anyone else who thinks there was some merit in the old models of intellectual property. For many college students, sharing files is considered an act of civil disobedience. That would mean

If we choose to pry culture away from capitalism while the rest of life is still capitalistic, culture will become a slum. In fact, online culture increasingly resembles a slum in disturbing ways. Slums have more advertising than wealthy neighborhoods, for instance. People are meaner in slums; mob rule and vigilantism are commonplace. If there is a trace of "slumming" in the way that many privileged young people embrace current online culture, it is perhaps an echo of 1960s counterculture.

that stealing digital material puts you in the company of Gandhi and Martin Luther King!*

It's true that the record companies have not helped themselves. They have made a public fuss about suing the most sympathetic people, snooped obnoxiously, and so on. Furthermore, there's a long history of sleaze, corruption, creative accounting, and price fixing in the music business.

Dreams Still Die Hard

By 2008, some of the leading lights of the open culture movement started to acknowledge the obvious, which is that not everyone has benefited from the movement. A decade ago we all assumed, or at least hoped, that the net would bring so many benefits to so many people that those unfortunates who weren't being paid for what they used to do would end up doing even better by finding new ways to get paid. You still hear that argument being made, as if people lived forever and can afford to wait an eternity to have the new source of wealth revealed to them.

Kevin Kelly wrote in 2008 that the new utopia

> is famously good news for two classes of people: a few lucky aggre-gators, such as Amazon and Netflix, and 6 billion consumers. Of those two, I think consumers earn the greater reward from the wealth hidden in infinite niches.
>
> But the long tail is a decidedly mixed blessing for creators. Individual artists, producers, inventors and makers are over-looked in the equation. The long tail does not raise the sales of cre-

---

*For an example of this common rationalization, here's a quote from an essay by "Sharkhead007" found on the site Big Nerds, which describes itself as a "free essay and coursework database" (meaning students use it to avoid writing assignments): "Critics would say that . . . if the government says something is illegal, it is morally wrong to go against it. However, Henry David Thoreau wrote a famous essay called Civil Disobedience, which described that sometimes the public has to revolt against law . . . Public activists and leaders such as Gandhi and Martin Luther King Jr. adopted the ideas expressed in Thoreau's essay and used them to better the lives of the people they were fighting for. Downloading music from the Internet, although it may not be as profound as freeing people from bondage and persecution, is a form of civil disobedience. It is a revolt against a corrupt system put in place for the sole purpose of making money, regardless of the welfare of the consumer or the artist."

*ators much, but it does add massive competition and endless
downward pressure on prices. Unless artists become a large aggregator of other artists' works, the long tail offers no path out of the
quiet doldrums of minuscule sales.*

The people who devote their lives to making committed cultural
expression that can be delivered through the cloud—as opposed to
casual contributions that require virtually no commitment—well, those
people are, Kevin acknowledges, the losers.

His new advice at the time was similar to the sorts of things we used
to suggest in fits of anticipation and wild hope ten, fifteen, and even
twenty-five years ago. He suggested that artists, musicians, or writers
find something that isn't digital related to their work, such as live
appearances, T-shirt sales, and so on, and convince a thousand people to
spend $100 each per year for whatever that is. Then an artist could earn
$100,000 a year.

I very much want to believe that this can be done by more than a tiny
number of people who happen to benefit from unusual circumstances.
The occasional dominatrix or life coach can use the internet to implement this plan. But after ten years of seeing many, many people try, I
fear that it won't work for the vast majority of journalists, musicians,
artists, and filmmakers who are staring into career oblivion because of
our failed digital idealism.

My skepticism didn't come easily. Initially I assumed that entrepreneurial fervor and ingenuity would find a way. As part of researching
this book, I set out once again to find some cultural types who were ben-
efiting from open culture.

The Search

We have a baseline in the form of the musical middle class that is being
put out of business by the net. We ought to at least find support in the
new economy for them. Can 26,000 musicians each find 1,000 true
fans? Or can 130,000 each find between 200 and 600 true fans? Furthermore, how long would be too long to wait for this to come about?
Thirty years? Three hundred years? Is there anything wrong with endur-

ing a few lost generations of musicians while we wait for the new solution to emerge?

The usual pattern one would expect is an S curve: there would be only a small number of early adaptors, but a noticeable trend of increase in their numbers. It is common in Silicon Valley to see incredibly fast adoption of new behaviors. There were only a few pioneer bloggers for a little while—then, suddenly, there were millions of them. The same could happen for musicians making a living in the new economy.

So at this point in time, a decade and a half after the start of the web, a decade after the widespread adoption of music file sharing, how many examples of musicians living by new rules should we expect to find?

Just to pick a rough number out of the air, it would be nice if there were 3,000 by now. Then maybe in a few years there would be 30,000. Then the S curve would manifest in full, and there would be 300,000. A new kind of professional musician ought to thunder onto the scene with the shocking speed of a new social networking website.

Based on the rhetoric about how much opportunity there is out there, you might think that looking for 3,000 is cynical. There must be tens of thousands already! Or you might be a realist, and think that it's still early; 300 might be a more realistic figure.

I was a little afraid to just post about my quest openly on the net, because even though I'm a critic of the open/free orthodoxy I didn't want to jinx it if it had a chance. Suppose I came up with a desultory result? Would that discourage people who would otherwise have made the push to make the new economy work?

Kevin Kelly thought my fear was ridiculous. He's more of a technological determinist: he thinks the technology will find a way to achieve its destiny whatever people think. So he volunteered to publicize my quest on his popular Technium blog in the expectation that exemplars of the new musical economy would come forward.

I also published a fire-breathing opinion piece in the *New York Times* and wrote about my fears in other visible places, all in the hope of inspiring contact from the new vanguard of musicians who are making a living off the open web.

In the old days—when I myself was signed to a label—there were a few major artists who made it on their own, like Ani DiFranco. She became a *millionaire* by selling her own CDs when they were still a high-margin product people were used to buying, back before the era of file sharing. Has a new army of Ani DiFrancos started to appear?

The Case of the Missing Beneficiaries

To my shock, I have had trouble finding even a handful of musicians who can be said to be following in DiFranco's footsteps. Quite a few musicians contacted me to claim victory in the new order, but again and again, they turned out to not be the real thing.

Here are some examples of careers that *do* exist but do not fill me with hope for the future:

▶ **The giant musical act from the old days of the record business, grabbing a few headlines by posting music for free downloading:** Radiohead is an example. I want to live in a world where new musicians can potentially succeed to the degree Radiohead has succeeded, but under a new order, not the old order. Where are they?

▶ **The aggregator:** A handful of musicians run websites that aggregate the music of hundreds or thousands of others. There are a few services that offer themed streaming music, for instance. One is a specialized new age music website that serves some paying yoga studios. The aggregator in this case is not Google, so only a trickle of money is made. The aggregated musicians make essentially nothing. Very few people can be aggregators, so this career path will not "scale," as we say in Silicon Valley.

▶ **The jingle/sound track/TV composer:** You can still make money from getting music placed in a setting that hasn't been destroyed by file sharing yet. Some examples are movie and TV sound tracks, commercial jingles, and so on. You can use internet presence to promote this kind of career. The problem with this strategy in the long term is that these paying options are themselves under siege.

▸ **The vanity career:** This is a devilish one. Music is glamorous, so there are perhaps more people who claim to be making a living as musicians than are actually doing so. There have probably always been way more people who have tried to have a music career than have succeeded at it. This is massively true online. There are hundreds of thousands of musicians seeking exposure on sites like MySpace, Bebo, YouTube, and on and on, and it is absolutely clear that most of them are not making a living from being there.

There is a seemingly limitless supply of people who want to pretend that they have professional music careers and will pay flacks to try to create the illusion. I am certainly not a private detective, but it takes only a few casual web searches to discover that a particular musician inherited a fortune and is barely referenced outside of his own website.

▸ **Kids in a van:** If you are young and childless, you can run around in a van to gigs, and you can promote those gigs online. You will make barely any money, but you can crash on couches and dine with fans you meet through the web. This is a good era for that kind of musical adventure. If I were in my twenties I would be doing it. But it is a youthiness career. Very few people can raise kids with that lifestyle. It's treacherous in the long run, as youth fades.

One example of success brought up again and again is Jonathan Coulton. He has nice career centered on spoofs and comedy songs, and his audience is the geeky crowd. He is certainly not becoming a millionaire, but at least he seems to have authentically reached the level of being able to reliably support a family without the assistance of the old-media model (though he does have a Hollywood agent, so he isn't an example to please the purist). There were only a handful of other candidates. The comedy blogger Ze Frank occasionally recorded tunes on his site, for example, and made money from a liquor ad placed there.

The tiny number of success stories is worrisome. The history of the web is filled with novelty-driven success stories that can never be repeated. One young woman started a website simply asking for donations to help her pay down her credit cards, and it worked! But none of the many people who tried to replicate her trick met with success.

This is astonishing to me. By now, a decade and a half into the web era, when iTunes has become the biggest music store, in a period when companies like Google are the beacons of Wall Street, shouldn't there at least be a few thousand initial pioneers of a new kind of musical career who can survive in our utopia? Maybe more will appear soon, but the current situation is discouraging.

Up-and-coming musicians in the open world can increasingly choose between only two options: they can try to follow the trail of mouse clicks laid down by Jonathan Coulton (and apparently almost no one can do that) or they can seek more reliable sustenance, by becoming refugees within the last dwindling pockets of the old-media world they were just assaulting a moment before.

The people who are perhaps the most screwed by open culture are the middle classes of intellectual and cultural creation. The freelance studio session musician faces diminished prospects, for instance. Another example, outside of the world of music, is the stringer selling reports to newspapers from a war zone. These are both crucial contributors to culture and democracy. Each pays painful dues and devotes years to honing a craft. They used to live off the trickle-down effects of the old system, and, like the middle class at large, they are precious. They get nothing from the new system.

Of course, eventually the situation might become transformed into something better. Maybe after a generation or two without professional musicians, some new habitat will emerge that will bring them back.

CHAPTER 6

THE LORDS OF THE CLOUDS RENOUNCE FREE WILL IN ORDER TO BECOME INFINITELY LUCKY

OUT-OF-CONTROL financial instruments are linked to the fates of musicians and the fallacies of cybernetic totalism.

Regional Fates

China's precipitous climb into wealth has been largely based on cheap, high-quality labor. But the real possibility exists that sometime in the next two decades a vast number of jobs in China and elsewhere will be made obsolete by advances in cheap robotics so quickly that it will be a cruel shock to hundreds of millions of people.

If waves of technological change bring new kinds of employment with them, what will it be like? Thus far, all computer-related technologies built by humans are endlessly confusing, buggy, tangled, fussy, and error-ridden. As a result, the icon of employment in the age of information has been the help desk.

For many years I've proposed that the "help desk," defined nobly and broadly to include such things as knowledge management, data forensics, software consulting, and so on, can provide us with a way to imagine a world in which capitalism and advanced technology can coexist with a fully employed population of human beings. This is a scenario I call "Planet of the Help Desks."

This brings us to India. India's economy has been soaring at the same time as China's, much to the amazement of observers everywhere, but on a model that is significantly different from China's. As Esther Dyson has pointed out, the Indian economy excels in "nonroutine" services.

India, thanks to its citizens' facility with English, hosts a huge chunk of the world's call centers, as well as a significant amount of software development, creative production like computer animation, outsourced administrative services, and, increasingly, health care.

America in Dreamland

Meanwhile, the United States has chosen a different path entirely. While there is a lot of talk about networks and emergence from the top American capitalists and technologists, in truth most of them are hoping to thrive by controlling the network that everyone else is forced to pass through.

Everyone wants to be a lord of a computing cloud. For instance, James Surowiecki in *The Wisdom of Crowds* extols an example in which an online crowd helped find gold in a gold mine even though the crowd didn't own the gold mine.

There are many forms of this style of yearning. The United States still has top universities and corporate labs, so we'd like the world to continue to accept intellectual property laws that send money our way based on our ideas, even when those ideas are acted on by others. We'd like to indefinitely run the world's search engines, computing clouds, advertising placement services, and social networks, even as our old friend/demon Moore's law makes it possible for new competitors to suddenly appear with ever greater speed and thrift.

We'd like to channel the world's finances through our currency to the benefit of our hedge fund schemes. Some of us would like the world to pay to watch our action movies and listen to our rock music into the indefinite future, even though others of us have been promoting free media services in order to own the cloud that places ads. Both camps are hoping that one way or another they will own the central nodes of the network even as they undermine each other.

Once again, this is an oversimplification. There are American factories and help desks. But, to mash up metaphors, can America maintain a virtual luxury yacht floating on the sea of the networks of the world? Or will our central tollbooth on all smart things sink under its own weight into an ocean of global connections? Even if we can win at the game, not many Americans will be employed keeping our yacht afloat, because it looks as though India will continue to get better at running help desks.

I'll be an optimist and suggest that America will somehow convince the world to allow us to maintain our privileged role. The admittedly flimsy reasons are that a) we've done it before, so they're used to us, and b) the alternatives are potentially less appealing to many global players, so there might be widespread grudging acceptance of at least some kinds of long-term American centrality as a least-bad option.

Computationally Enhanced Corruption

Corruption has always been possible without computers, but computers have made it easier for criminals to pretend even to themselves that they are not aware of their own schemes. The savings and loan scandals of the 1980s were possible without extensive computer network services. All that was required was a misuse of a government safety net. More recent examples of cataclysmic financial mismanagement, starting with Enron and Long-Term Capital Management, could have been possible only with the use of big computer networks. The wave of financial calamities that took place in 2008 were significantly cloud based.

No one in the pre–digital cloud era had the mental capacity to lie to him- or herself in the way we routinely are able to now. The limitations of organic human memory and calculation used to put a cap on the intricacies of self-delusion. In finance, the rise of computer-assisted hedge funds and similar operations has turned capitalism into a search engine. You tend the engine in the computing cloud, and it searches for money. It's analogous to someone showing up in a casino with a supercomputer and a bunch of fancy sensors. You can certainly win at gambling with high-tech help, but to do so you must supercede the game you are pretending to play. The casino will object, and in the case of investment in the real world, society should also object.

Visiting the offices of financial cloud engines (like high-tech hedge funds) feels like visiting the Googleplex. There are software engineers all around, but few of the sorts of topical experts and analysts who usually populate investment houses. These pioneers have brought capitalism into a new phase, and I don't think it's working.

In the past, an investor had to be able to understand at least something about what an investment would actually accomplish. Maybe a building would be built, or a product would be shipped somewhere, for instance. No more. There are so many layers of abstraction between the new kind of elite investor and actual events on the ground that the investor no longer has any concept of what is actually being done as a result of investments.

The Cloudy Edge Between Self-Delusion and Corruption

True believers in the hive mind seem to think that no number of layers of abstraction in a financial system can dull the efficacy of the system. According to the new ideology, which is a blending of cyber-cloud faith and neo–Milton Friedman economics, the market will not only do what's best, it will do better the less people understand it. I disagree. The financial crisis brought about by the U.S. mortgage meltdown of 2008 was a case of too many people believing in the cloud too much.

Each layer of digital abstraction, no matter how well it is crafted, contributes some degree of error and obfuscation. No abstraction corresponds to reality perfectly. A lot of such layers become a system unto themselves, one that functions apart from the reality that is obscured far below. Making money in the cloud doesn't necessarily bring rain to the ground.

The Big *N*

Here we come to one way that the ideal of "free" music and the corruption of the financial world are connected.

Silicon Valley has actively proselytized Wall Street to buy into the doctrines of open/free culture and crowdsourcing. According to Chris Anderson, for instance, Bear Stearns issued a report in 2007 "to address *pushback* and other objections from media industry heavyweights who make up a big part of Bear Stearns's client base."

What the heavyweights were pushing back against was the Silicon

Valley assertion that "content" from identifiable humans would no longer matter, and that the chattering of the crowd with itself was a better business bet than paying people to make movies, books, and music.

Chris identified his favorite quote from the Bear Stearns report:

> *For as long as most can recall, the entertainment industry has lived by the axiom "content is king." However, no one company has proven consistently capable of producing "great content," as evidenced by volatility in TV ratings and box office per film for movie studios, given the inherent fickleness of consumer demand for entertainment goods.*

As Chris explains, "despite the bluster about track records and taste . . . it's all a crapshoot. Better to play the big-n statistical game of User Generated Content, as YouTube has, than place big bets on a few horses like network TV."

"Big-n" refers to "n," a typical symbol for a mathematical variable. If you have a giant social network, like Facebook, perhaps some variable called n gains a big value. As n gets larger, statistics become more reliable. This might also mean, for example, that it becomes more likely that someone in the crowd will happen to provide you with a free gem of a song or video.

However, it must be pointed out that in practice, even if you believe in the big n as a substitute for judgment, n is almost never big enough to mean anything on the internet. As vast as the internet has become, it usually isn't vast enough to generate valid statistics. The overwhelming majority of entries garnering reviews on sites like Yelp or Amazon have far too few reviewers to reach any meaningful level of statistical utility. Even when n is large, there's no guarantee it's valid.

In the old order, there were occasional smirks and groans elicited by egregious cases of incompetence. Such affronts were treated as exceptions to the rule. In general it was assumed that the studio head, the hedge fund manager, and the CEO actually did have some special skills, some reason to be in a position of great responsibility.

In the new order, there is no such presumption. The crowd works for free, and statistical algorithms supposedly take the risk out of making bets if you are a lord of the cloud. Without risk, there is no need for skill. But who is that lord who owns the cloud that connects the crowd? Not just anybody. A lucky few (for luck is all that can possibly be involved) will own it. Entitlement has achieved its singularity and become infinite.

Unless the algorithm actually isn't perfect. But we're rich enough that we can delay finding out if it's perfect or not. This is the grand unified scam of the new ideology.

It should be clear that the madness that has infected Wall Street is just another aspect of the madness that insists that if music *can* be delivered for free, it *must* be delivered for free. The Facebook Kid and the Cloud Lord are serf and king of the new order.

In each case, human creativity and understanding, especially one's own creativity and understanding, are treated as worthless. Instead, one trusts in the crowd, in the big *n*, in the algorithms that remove the risks of creativity in ways too sophisticated for any mere person to understand.

CHAPTER 7

THE PROSPECTS FOR HUMANISTIC CLOUD ECONOMICS

ALTERNATIVES ARE PRESENTED to doctrinaire ideas about digital economics.

The Digital Economy: First Thought, Best Thought

A natural question to ask at this point is, Are there any alternatives, any options, that exist apart from the opposing poles of old media and open culture?

Early on, one of the signal ideas about how a culture with a digital network could—and should—work was that the need for money might be eliminated, since such a network could keep track of fractional barters between very large groups of people. Whether that idea will ever come back into the discussion I don't know, but for the foreseeable future we seem to be committed to using money for rent, food, and medicine. So is there any way to bring money and capitalism into an era of technological abundance without impoverishing almost everyone? One smart idea came from Ted Nelson.

Nelson is perhaps the most formative figure in the development of online culture. He invented the digital media link and other core ideas of connected online media back in the 1960s. He called it "hypermedia."

Nelson's ambitions for the economics of linking were more profound than those in vogue today. He proposed that instead of copying digital media, we should effectively keep only one copy of each cultural expression—as with a book or a song—and pay the author of that expression a small, affordable amount whenever it is accessed. (Of course, as a matter of engineering practice, there would have to be many copies in order for the system to function efficiently, but that would be an internal detail, unrelated to a user's experience.)

As a result, anyone might be able to get rich from creative work. The people who make a momentarily popular prank video clip might earn a lot of money in a single day, but an obscure scholar might eventually earn as much over many years as her work is repeatedly referenced. But note that this is a very different idea from the long tail, because it rewards individuals instead of cloud owners.

The popularity of amateur content today provides an answer to one of the old objections to Nelson's ideas. It was once a common concern that most people would not want to be creative or expressive, ensuring that only a few artists would get rich and that everyone else would starve. At one event, I remember Nelson trying to speak and young American Maoists shouting him down because they worried that his system would favor the intellectual over the peasant.

I used to face this objection constantly when I talked about virtual reality (which I discuss more fully in Chapter 14). Many a lecture I gave in the 1980s would end with a skeptic in the audience pointing out loudly and confidently that only a tiny minority of people would ever write anything online for others to read. They didn't believe a world with millions of active voices was remotely possible—but that is the world that has come to be.

If we idealists had only been able to convince those skeptics, we might have entered into a different, and better, world once it became clear that the majority of people are indeed interested in and capable of being expressive in the digital realm.

Someday I hope there will be a genuinely universal system along the lines proposed by Nelson. I believe most people would embrace a social contract in which bits have value instead of being free. Everyone would have easy access to everyone else's creative bits at reasonable prices—and everyone would get paid for their bits. This arrangement would celebrate personhood in full, because personal expression would be valued.

Pick Your Poison

There is an intensely strong libertarian bias in digital culture—and what I have said in the preceding section is likely to enrage adherents of digital libertarianism.

It's not hard to see why. If I'm suggesting a universal system, inspired by Ted Nelson's early work, doesn't that mean the government is going to get in the middle of your flow of bits in order to enforce laws related to compensation for artists? Wouldn't that be intrusive? Wouldn't it amount to a loss of liberty?

From the orthodox point of view, that's how it probably looks, but I hope to persuade even the truest believers that they have to pick their poison—and that the poison I'm suggesting here is ultimately preferable, *especially* from a libertarian perspective.

It's important to remember the extreme degree to which we make everything up in digital systems, at least during the idyllic period before lock-in constricts our freedoms. Today there is still time to reconsider the way we think about bits online, and therefore we ought to think hard about whether what will otherwise become the official future is really the best we can do.

Let's take money—the original abstract information system for managing human affairs—as an example. It might be tempting to print your own money, or, if you're the government, to print an excessive amount of it. And yet smart people choose not to do either of these things.

> The scarcity of money, as we know it today, is artificial, but everything about information is artificial. Without a degree of imposed scarcity, money would be valueless.

It is a common assertion that if you copy a digital music file, you haven't destroyed the original, so nothing was stolen. The same thing could be said if you hacked into a bank and just added money to your online account. (Or, for that matter, when traders in exotic securities made bets on stupendous transactions of arbitrary magnitudes, leading to the global economic meltdown in 2008.) The problem in each case is not that you stole from a specific person but that you undermined the artificial scarcities that allow the economy to function. In the same way, creative expression on the internet will benefit from a social contract that imposes a *modest* degree of artificial scarcity on information.

In Ted Nelson's system, there would be no copies, so the idea of copy protection would be mooted. The troubled idea of digital rights management—that cumbersome system under which you own a copy of bits you bought, but not really, because they are still managed by the seller—would not exist. Instead of collections of bits being offered as a product, they would be rendered as a service.

Creative expression could then become the most valuable resource in a future world of material abundance created through the triumphs of technologists. In my early rhetoric about virtual reality back in the 1980s, I always said that in a virtual world of infinite abundance, only creativity could ever be in short supply—thereby ensuring that creativity would become the most valuable thing.

Recall the earlier discussion of Maslow's hierarchy. Even if a robot that maintains your health will only cost a penny in some advanced future, how will you earn that penny? Manual labor will be unpaid, since cheap robots will do it. In the open culture future, your creativity and expression would also be unpaid, since you would be a volunteer in the army of the long tail. That would leave nothing for you.

Everything Sounds Fresh When It Goes Digital—Maybe Even Socialism

The only alternative to some version of Nelson's vision in the long run— once technology fulfills its potential to make life easy for everyone— would be to establish a form of socialism.

Indeed, that was the outcome that many foresaw. Maybe socialism can be made compassionate and efficient (or so some digital pioneers daydreamed) if you just add a digital backbone.

I am not entirely dismissive of the prospect. Maybe there is a way it can be made to work. However, there are some cautions that I hope any new generations of digital socialists will take to heart.

A sudden advent of socialism, just after everyone has slid down Maslow's pyramid into the mud, is likely to be dangerous. The wrong people often take over when a revolution happens suddenly. (See: Iran.) So if socialism is where we are headed, we ought to be talking about it now so that we can approach it incrementally. If it's too toxic a subject to even talk about openly, then we ought to admit we don't have the abilities to deal with it competently.

I can imagine that this must sound like a strange exhortation to some readers, since socialism might seem to be the ultimate taboo in libertarian Silicon Valley, but there is an awful lot of stealth socialism going on beneath the breath in digital circles. This is particularly true for young people whose experience of markets has been dominated by the market failures of the Bush years.

It isn't crazy to imagine that there will be all sorts of new, vast examples of communal cooperation enabled through the internet. The initial growth of the web itself was one, and even though I don't like the way people are treated in web 2.0 designs, they have provided many more examples.

A prominent strain of enthusiasm for wikis, long tails, hive minds, and so on incorporates the presumption that one profession after another will be demonetized. Digitally connected mobs will perform more and more services on a collective volunteer basis, from medicine to solving crimes, until all jobs are done that way. The cloud lords might still be able to hold on to their thrones—which is why even the most ardent Silicon Valley capitalists sometimes encourage this way of thinking.

This trajectory begs the question of how a person who is volunteering for the hive all day long will earn rent money. Will living space become something doled out by the hive? (Would you do it with Wikipedia-style edit wars or Digg-style voting? Or would living space only be inherited, so that your station in life was predetermined? Or would it be allocated at random, reducing the status of free will?)

Digital socialists must avoid the trap of believing that a technological makeover has solved *all* the problems of socialism just because it can solve *some* of them. Getting people to cooperate is not enough.

Private property in a market framework provides one way to avoid a deadening standard in shaping the boundaries of privacy. This is why a market economy can enhance individuality, self-determination, and dignity, at least for those who do well in it. (That not everybody does well is a problem, of course, and later on I'll propose some ways digital tech might help with that.)

Can a digital version of socialism also provide dignity and privacy? I view that as an important issue—and a very hard one to resolve.

It Isn't Too Late

How, exactly, could a transition from open copying to paid access work? This is a situation in which there need to be universal, governmental solutions to certain problems.

People have to all agree in order for something to have monetary value. For example, if everyone else thinks the air is free, it's not going to be easy to convince me to start paying for it on my own. These days it amazes me to remember that I once purchased enough music CDs to fill a wall of shelves—but it made sense at the time, because everyone I knew also spent a lot of money on them.

Perceptions of fairness and social norms can support or undermine any economic idea. If I know my neighbor is getting music, or cable TV, or whatever, for free, it becomes a little harder to get me to pay for the same things.* So for that reason, if all of us are to earn a living when the machines get good, we will have to agree that it is worth paying for one another's elevated cultural and creative expressions.

There are other cases where consensus will be needed. One online requirement that hurt newspapers before they gave up and went "open" was the demand that you enter your password (and sometimes your new credit card numbers) on each and every paid site that you were interested in accessing. You could spend every waking minute entering such information in a world of millions of wonderful paid-content sites. There has to be a universal, simple system. Despite some attempts, it doesn't look as if the industry is able to agree on how to make this happen, so this annoyance seems to define a natural role for government.

It is strange to have to point this out, but given the hyper-libertarian atmosphere of Silicon Valley, it's important to note that government isn't always bad. I like the "Do not call" list, for instance, since it has contained the scourge of telemarketing. I'm also glad we only have one currency, one court system, and one military. Even the most extreme libertarian must admit that fluid commerce has to flow through channels that amount to government.

---

*This principle has even been demonstrated in dogs and monkeys. When Dr. Friederike Range of the University of Vienna allowed dogs in a test to see other dogs receive better rewards, jealousy ensued. Dogs demand equal treatment in order to be trained well. Frans de Waal at Emory University found similar results in experiments with capuchin monkeys.

Of course, one of the main reasons that digital entrepreneurs have tended to prefer free content is that it costs money to manage micropayments. What if it costs you a penny to manage a one-penny transaction? Any vendor who takes on the expense is put at a disadvantage.

In such a case, the extra cost should be borne by the whole polis, as a government function. That extra penny isn't wasted—it's the cost of maintaining a social contract. We routinely spend more money incarcerating a thief than the thief stole in the first place. You could argue that it would be cheaper to not prosecute small crimes and just reimburse the victims. But the reason to enforce laws is to create a livable environment for everyone. It's exactly the same with putting value on individual human creativity in a technologically advanced world.

We never record the true cost of the existence of money because most of us put in volunteer time to maintain the social contract that gives money its value. No one pays you for the time you take every day to make sure you have cash in your wallet, or to pay your bills—or for the time you spend worrying about the stuff. If that time were reimbursed, then money would become too expensive as a tool for a society.

In the same way, the maintenance of the liberties of capitalism in a digital future will require a general acceptance of a social contract. We will pay a tax to have the ability to earn money from our creativity, expression, and perspective. It will be a good deal.

The Transition

The transition would not have to be simultaneous and universal, even though the ultimate goal would be to achieve universality. One fine day your ISP could offer you an option: You could stop paying your monthly access charge in exchange for signing up for the new social contract in which you pay for bits. If you accessed no paid bits in a given month, you would pay nothing for that month.

If you chose to switch, you would have the potential to earn money from your bits—such as photos and music—when they were visited by other people. You'd also pay when you visited the bits of others. The total you paid per month would, on average, initially work out to be similar to what you paid before, because that is what the market would bear. Gradually, more and more people would make the transition, because people are entrepreneurial and would like the chance to try to make money from their bits.

The details would be tricky—but certainly no more so than they are in the current system.

What Makes Liberty Different from Anarchy Is Biological Realism

The open culture crowd believes that human behavior can only be modified through involuntary means. This makes sense for them, because they aren't great believers in free will or personhood.

For instance, it is often claimed by open culture types that if you can't make a perfect copy-protection technology, then copy prohibitions are pointless. And from a technological point of view, it is true that you can't make a perfect copy-protection scheme. If flawless behavior restraints are the only potential influences on behavior in a case such as this, we might as well not ask anyone to ever pay for music or journalism again. According to this logic, the very idea is a lost cause.

But that's an unrealistically pessimistic way of thinking about people. We have already demonstrated that we're better than that. It's easy to break into physical cars and houses, for instance, and yet few people do so. Locks are only amulets of inconvenience that remind us of a social contract we ultimately benefit from. It is only human choice that makes the human world function. Technology can motivate human choice, but not replace it.

I had an epiphany once that I wish I could stimulate in everyone else. The plausibility of our human world, the fact that the buildings don't all fall down and you can eat unpoisoned food that someone grew, is immediate palpable evidence of an ocean of goodwill and good behavior from almost everyone, living or dead. We are bathed in what can be called love.

And yet that love shows itself best through the constraints of civilization, because those constraints compensate for the flaws of human nature. We must see ourselves honestly, and engage ourselves realistically, in order to become better.

CHAPTER 8

THREE POSSIBLE FUTURE
DIRECTIONS

IN THIS CHAPTER, I will discuss three long-term projects that I have worked on in an effort to correct some of the problems I described in Chapter 4. I don't know for sure that any of my specific efforts to ensure that the digital revolution will enhance humanism rather than restrict it will work. But at the very least, I believe they demonstrate that the range of possible futures is broader than you might think if you listen only to the rhetoric of web 2.0 people.

Two of the ideas, telegigging and songles, address problems with the future of paid cultural expression. The third idea, formal financial expression, represents an approach to keeping the hive from ruining finance.

Telegigging

There was a time, before movies were invented, when live stage shows offered the highest production values of any form of human expression.

If canned content becomes a harder product to sell in the internet era, the return of live performance—in a new technological context—might be the starting point for new kinds of successful business plans.

Let's approach this idea first by thinking small. What if you could hire a live musician for a party, even if that musician was at a distance? The performance might feel "present" in your house if you had immersive, "holographic" projectors in your living room. Imagine telepresent actors, orators, puppeteers, and dancers delivering real-time interactive shows that include special effects and production values surpassing those of today's most expensive movies. For instance, a puppeteer for a child's birthday party might take children on a magical journey through a unique immersive fantasy world designed by the performer.

This design would provide performers with an offering that could be delivered reasonably because they wouldn't have to travel. Telepresent performance would also provide a value to customers that file sharing could not offer. It would be immune to the problems of online commerce that have shriveled the music labels.

Here we might finally have a scenario that could solve the problem of how musicians can earn a living online. Obviously, the idea of "teleperformance for hire" remains speculative at this time, but the technology appears to be moving in a direction that will make it possible.

Now let's think big. Suppose big stars and big-budget virtual sets, and big production values in every way, were harnessed to create a simulated world that home participants could enter in large numbers. This would be something like a cross between Second Life and teleimmersion.

In many ways this sort of support for a mass fantasy is what digital technology seems to be converging on. It is the vision many of us had in mind decades ago, in much earlier phases of our adventures as technologists. Artists and media entrepreneurs might evolve to take on new roles, providing the giant dream machine foreseen in a thousand science fiction stories.

Songles

A songle is a dongle for a song. A dongle is a little piece of hardware that you plug into a computer to run a piece of commercial software. It's like a physical key you have to buy in order to make the software work. It creates artificial scarcity for the software.

All the tchotchkes of the world—the coffee mugs, the bracelets, the nose rings—would serve double duty as keys to content like music.

There's a green angle here. All the schemes that presently succeed in getting people to pay for content involve the manufacture of extra hardware that would not otherwise be needed. These include music players such as iPods, cable TV boxes, gaming consoles, and so on. If people paid for content, there would be no need for these devices, since commonplace computer chips and displays would be good enough to perform all these tasks.

Songles would provide a physical approach to creating artificial scarcity. It might be less difficult to make the transition to songles than it would be to implement a more abstract approach to bringing expression back under the tent of capitalism.

You might wear a special necklace songle to a party, and music enabled by the necklace would come on automatically after you arrived, emanating from the entertainment system that is already providing the party with music. The necklace communicates with the entertainment system in order to make this happen. The musical mix at an event might be determined by the sum of the songles worn by everyone who shows up.

WHY BRING PHYSICAL OBJECTS BACK INTO MUSIC DISTRIBUTION?

▶ **To make the music business more romantic:** That's not just an enhancement; it's the central issue. Romance, in the broadest sense, is the product the music business sells. Contracts and credit card numbers are not romantic.

▶ **To lower the cost of promotion:** Music production and distribution costs have become low, but promotion costs are limitless. Since a songle is an object instead of a contract, its value is determined by the marketplace and can vary over time, even if traded informally. In order to be effective, songles must come in limited editions. This means that a songle can be an object for speculative investment. A fan who takes the trouble to listen to obscure new bands might benefit from having speculated on buying some of the bands' songles when they were unknown. Songles harness the psychology that makes lottery tickets sell to get people to listen to new music acts. Even better: once a person buys a songle, she is motivated to join in promoting its music, because she now has a stake in it.

WHAT WILL MONEY BE?

▶ **To broaden the channels by which music is sold and share promotion costs with players in those channels:** High-end, rare songles can be sold as accessories at fashion stores, while low-end songles might come bundled with a six-pack. Coffee mugs, sneakers, toothbrushes, dog collars, pens, and sunglasses would all make fine songles.

▶ **To raise the margin for high-prestige but low-volume (in the business sense!) music:** The stupidest thing among many stupid things in the music business is that the product always costs about the same even when a market segment would naturally choose a higher price if it were allowed to do so. For instance, a well-heeled opera fan pays about the same for a CD or a download as does a teenager listening to a teen idol of the moment. Songles for opera or fine jazz would be made by craftsmen from fine materials in much more limited editions. They would be expensive. Low-end songles would be manufactured by the same channel that provides toys. An increasing number of consumer items that might become songles these days have radio-frequency identification anyway, so there would be no additional manufacturing expense. Expensive limited-edition songles would probably accompany the introduction of new forms of pop music—in parallel with cheap large-volume editions—because there would be a fabulous market for them.

Formal Financial Expression*

Unlike the previous two sections, this one addresses the problems of the lords of the clouds, not the peasants.

One of the toughest problems we'll face as we emerge from the financial crisis that beset us in 2008 is that financiers ought to continue to

*Some of my collaborators in this research include Paul Borrill, Jim Herriot, Stuart Kauffman, Bruce Sawhill, Lee Smolin, and Eric Weinstein.

innovate in creating new financial instruments, even though some of them recently failed catastrophically doing just that. We need them to learn to do their job more effectively—and safely—in the future.

This is a crucial issue for our green future. As the world becomes more complex, we'll need innovative financial structures to manage new and unforeseen challenges. How do you finance massive conversions to green technologies that are partially centralized and partially decentralized? How can a financial design avoid catastrophic losses, as massive portions of the infrastructure of the old energy cycle are made obsolete? Battling global warming will require new patterns of development that in turn require new financial instruments.

However, it might be a while before governments allow much in the way of deep innovation in finance. Regulators were unable to keep up with some of the recent inventions; indeed, it is becoming sadly clear that in some cases the very people who invented financial instruments did not really understand them.

So this is our dilemma: How do we avoid putting a lid on innovation in finance after a huge crisis in confidence?

Economics is about how to best mix a set of rules we cannot change with rules that we can change. The rules we cannot change come from math and the state of physical reality at a given time (including such factors as the supply of natural resources). We hope the rules we can change will help us achieve the best results from those we can't. That is the rational side of economics.

But there is an irrational side to all human quests. Irrationality in a market is found not only in individuals, but in the economists who study them and in the regulators who attempt to steer their actions.

Sometimes people decide to continue to use a technology that disappoints again and again, even one that is deadly dangerous. Cars are a great example. Car accidents kill more people than wars, and yet we love cars.

Capitalism is like that. It gives us the buzz of freedom. We adore it even though it has crashed on occasion. We always pretend it will be the other person who is hurt.

Our willingness to suffer for the sake of the perception of freedom is remarkable. We believe in the bits housed in the computers of the financial world enough to continue to live by them, even when they sting us, because those bits, those dollars, are the abstractions that help us feel free.

Engineers sometimes take on the inherently absurd task of making a deliberately imperfect technology slightly less imperfect. For example, cars are usually designed to reach ridiculous, illegal speeds, because that makes us feel free—and in addition, they come with air bags. This is the absurdity of engineering for the real world.

So the task at hand has an unavoidably absurd quality. If economic engineering succeeds too well, the whole system could lose its appeal. Investors want to periodically feel that they are getting away with something, living on the edge, taking outlandish risks. We want our capitalism to feel wild, like a jungle, or like our most brilliant models of complex systems. Perhaps, though, we can find a way to keep the feeling while taming the system a bit.

One idea I'm contemplating is to use so-called AI techniques to create formal versions of certain complicated or innovative contracts that define financial instruments. Were this idea to take hold, we could sort financial contracts into two domains. Most transactions would continue to be described traditionally. If a transaction followed a cookie-cutter design, then it would be handled just as it is now. Thus, for instance, the sale of stocks would continue as it always has. There are good things about highly regular financial instruments: they can be traded on an exchange, for instance, because they are comparable.

But highly inventive contracts, such as leveraged default swaps or schemes based on high-frequency trades, would be created in an entirely new way. They would be denied ambiguity. They would be formally described. Financial invention would take place within the simplified logical world that engineers rely on to create computing-chip logic.

Reducing the power of expression of unconventional financial contracts might sound like a loss of fun for the people who invent them, but, actually, they will enjoy heightened powers. The reduction in flexibility doesn't preclude creative, unusual ideas at all. Think of all the varied chips that have been designed.

Constrained, formal systems can, in some cases, be analyzed in ways that more casual expressions cannot. This means that tools can be created to help financiers understand what they are doing with far more insight than was possible before. Once enhanced analytical strategies are possible, then financiers, regulators, and other stakeholders wouldn't have to rely solely on bottom-up simulation to examine the implications of what they are doing.

This premise has proven controversial. Technically inclined people who are enthusiasts for ideas related to "complexity" often want financial instruments to benefit from the same open qualities that define life, freedom, democracy, the law, language, poetry, and so on. Then there's an opposing camp of shell-shocked people who, because of our recent financial woes, want to clamp down and force finance into easy-to-regulate repetitive structures.

The economy is a tool, and there's no reason it has to be as open and wild as the many open and wild things of our experience. But it also doesn't have to be as tied down as some might want. It can and should have an intermediate level of complexity.

Formal financial expression would define an intermediate zone, which is not as open as life or democracy but not as closed as a public securities exchange. The structures in this zone could still be interesting, but they, and their composites, could also still be subject to certain formal analyses.

Would financiers accept such a development? At first it sounds like a limitation, but the trade-offs would turn out to be favorable to the entrepreneurial and experimental spirit.

There would be one standard formal representation of transactions, but also an open diversity of applications that make use of it. That means that financial designs would not have to follow preexisting contours and could be developed in a wide variety of ways, but could still be registered with regulators. The ability to register complex, creative ideas in a standard form would transform the nature of finance and its regulation. It would become possible to create a confidential, anonymous-except-by-court-order method for regulators to track unusual transactions. That would solve one huge recent problem, which was the impossibility of tallying a full accounting of how deep the hole was after the crash, since the exotic financial instruments were described in terms that could be subject to varying interpretations.

The ability to understand the implications of a wide range of innovative, nonstandard transactions will make it possible for central banks and other authorities to set policy in the future with a full comprehension of what they are doing. And *that* will allow financiers to be innovative. Without some method of eliminating the kind of institutional blindness that led to our recent financial catastrophes, it is hard to imagine how innovation in the financial sector will be welcomed again.

A cooperative international body would probably have specific requirements for the formal representation, but any individual application making use of it could be created by a government, a nongovernmental organization, an individual, a school, or a for-profit company. The formal transaction-representation format would be nonproprietary, but there would be a huge market for proprietary tools that make it useful. These tools would quickly become part of the standard practice of finance.

There would be a diversity of apps for *creating* contracts as well as analyzing them. Some would look like specialized word processors that create the illusion of writing a traditional contract, while others might have experimental graphic user interfaces. Instead of solely outputting a written contract of the usual sort to define a financial instrument, the parties would also generate an additional computer file that would be derived from a contract as part of the guided process of writing it. This file would define the structure of the financial instrument in the formal, internationally standardized way.

Applications analogous to Mathematica could be created that would transform, combine, simulate, and analyze transactions defined in these files.

For example:

▶ A given transaction could be restated from the point of view of a customer, a third party defining derivatives of it, a regulator, or other parties.

▶ It could also be analyzed within the curved space of an expanding or contracting economy (hopefully encouraging the correction of how granularities—which usually assume a static environment—are defined).

▶ The temporal aspects of the transaction could be analyzed so that indexes and other measurements could be tweaked to avoid artifacts due to inappropriate granularity.

▶ A transaction design could be input into simulations of a wide variety of scenarios to help analysts assess risks.

▶ Regulations could be expressed in a more general and abstract way. For instance, if a regulator became curious about whether a particular derivative should be understood as a form of insurance—which should only be allowed if the insurer has adequate reserves—it would be easy to make the necessary analysis. (This function would have prevented much of the current mess.)

▶ It should also be possible to detect the potential emergence of Ponzi schemes and the like within complex networks of transactions that might otherwise fool even those who designed them.

▶ Visualizations or other nonstandard presentations of transactions that would help legislators and other nonspecialists understand new ideas in transactions might be developed.

▶ A tool to help consumers cope with the monetary world might well come from an enlightened NGO or a university. I would hope to see foundations offering prizes for the best visualization, teaching, or planning tools for ordinary people, for instance.

This is an extremely ambitious vision, because, among other things, it involves the representation of ideas that are usually expressed in natural language (in contracts), and because, at the cloud level, it must reconcile multiple contracts that may often be underspecified and reveal ambiguities and/or contradictions in an emerging system of expressions.

But while these problems will be a headache for software developers, they might also ultimately force financiers to become better at describing what they do. They aren't artists who should be allowed to make ambiguous, impossible-to-parse creations. The need to interoperate more tightly with the "dumbness" of software could help them undertake their work more clearly and safely.

Furthermore, this sort of transaction representation has already been done internally within some of the more sophisticated hedge funds. Computer science is mature enough to take this problem on.

PART THREE

THE UNBEARABLE
THINNESS OF FLATNESS

THREE WARNINGS have been presented in the previous chapters, conveying my belief that cybernetic totalism will ultimately be bad for spirituality, morality, and business. In my view, people have often respected bits too much, resulting in a creeping degradation of their own qualities as human beings.

This section addresses another kind of danger that can arise from believing in bits too much. Recall that in Chapter 1 I made a distinction between ideal and real computers. Ideal computers can be experienced when you write a small program. They seem to offer infinite possibilities and an extraordinary sense of freedom. Real computers are experienced when we deal with large programs. They can trap us in tangles of code and make us slaves to legacy—and not just in matters of obscure technological decisions. Real computers reify our philosophies through the process of lock-in before we are ready.

People who use metaphors drawn from computation when they think about reality naturally prefer to think about ideal computers instead of real ones. Thus, the cultural software engineers usually present us with a world in which each cultural expression is like a brand-new tiny program, free to be anything at all.

That's a sweet thought, but it brings about an unfortunate side effect. If each cultural expression is a brand-new tiny program, then they are all aligned on the same starting line. Each one is created using the same resources as every other one.

This is what I call a "flat" global structure. It suggests a happy world to software technologists, because every little program in a flat global structure is born fresh, offering a renewing whiff of the freedom of tiny code.

Software people know that it's useless to continue to write tiny programs forever. To do anything useful, you have to take the painful plunge into large code. But they seem to imagine that the domain of tiny, virginal expression is still going to be valid in the spheres of culture and, as I'll explain, science.

That's one reason the web 2.0 designs strongly favor flatness in cultural expression. But I believe that flatness, as applied to human affairs, leads to blandness and meaninglessness. And there are analogous problems related to the increasing popularity of flatness in scientific thought. When applied to science, flatness can cause confusion between methodology and expression.

CHAPTER 9

ReTROPOLIS

AN ANOMALY IN popular music trends is examined.

Second-Order Culture

What's gone so stale with internet culture that a batch of tired rhetoric from my old circle of friends has become sacrosanct? Why can't anyone younger dump our old ideas for something original? I long to be shocked and made obsolete by new generations of digital culture, but instead I am being tortured by repetition and boredom.

For example: the pinnacle of achievement of the open software movement has been the creation of Linux, a derivative of UNIX, an old operating system from the 1970s. Similarly, the less techie side of the open culture movement celebrates the creation of Wikipedia, which is a copy of something that already existed: an encyclopedia.

What I'm saying here is independent of whether the typical claims made by web 2.0 and wiki enthusiasts are true. Let's just stipulate for the sake of argument that Linux is as stable and secure as any historical derivative of UNIX and that Wikipedia is as reliable as other encyclopedias. It's still strange that generations of young, energetic, idealistic people would perceive such intense value in creating them.

Let's suppose that back in the 1980s I had said, "In a quarter century, when the digital revolution has made great

> There's a rule of thumb you can count on in each succeeding version of the web 2.0 movement: the more radical an online social experiment is claimed to be, the more conservative, nostalgic, and familiar the result will actually be.

progress and computer chips are *millions* of times faster than they are now, humanity will finally win the prize of being able to write a new encyclopedia and a new version of UNIX!" It would have sounded utterly pathetic.

The distinction between first-order expression and derivative expression is lost on true believers in the hive. First-order expression is when someone presents a whole, a work that integrates its own worldview and aesthetic. It is something genuinely new in the world.

Second-order expression is made of fragmentary reactions to first-order expression. A movie like *Blade Runner* is first-order expression, as was the novel that inspired it, but a mashup in which a scene from the movie is accompanied by the anonymous masher's favorite song is not in the same league.

I don't claim I can build a meter to detect precisely where the boundary between first- and second-order expression lies. I *am* claiming, however, that the web 2.0 designs spin out gobs of the latter and choke off the former.

It is astonishing how much of the chatter online is driven by fan responses to expression that was originally created within the sphere of old media and that is now being destroyed by the net. Comments about TV shows, major movies, commercial music releases, and video games must be responsible for almost as much bit traffic as porn. There is certainly nothing wrong with that, but since the web is killing the old media, we face a situation in which culture is effectively eating its own seed stock.

Schlock Defended

The more original material that does exist on the open net is all too often like the lowest-production-cost material from the besieged, old-fashioned, copy-written world. It's an endless parade of "News of the Weird," "Stupid Pet Tricks," and *America's Funniest Home Videos*.

This is the sort of stuff you'll be directed to by aggregation services like YouTube or Digg. (That, and endless propaganda about the merits of open culture. Some stupefying, dull release of a version of Linux will usually be a top world headline.)

I am not being a snob about this material. I like it myself once in a while. Only people can make schlock, after all. A bird can't be schlocky when it sings, but a person can. So we can take existential pride in schlock. All I am saying is that we already had, in the predigital world, all the kinds of schlock you now find on the net. Making echoes of this material in the radical, new, "open" world accomplishes nothing. The cumulative result is that online culture is fixated on the world as it was before the web was born.

By most estimates, about half the bits coursing through the internet originated as television, movie, or other traditional commercial content, though it is difficult to come up with a precise accounting.

BitTorrent, a company that maintains only one of the many protocols for delivering such content, has at times claimed that its users alone are taking up more than half of the bandwidth of the internet. (BitTorrent is used for a variety of content, but a primary motivation to use it is that it is suitable for distributing large files, such as television shows and feature-length movies.)

The internet was, of course, originally conceived during the Cold War to be capable of surviving a nuclear attack. Parts of it can be destroyed without destroying the whole, but that also means that parts can be known without knowing the whole. The core idea is called "packet switching."

A packet is a tiny portion of a file that is passed between nodes on the internet in the way a baton is passed between runners in a relay race. The packet has a destination address. If a particular node fails to acknowledge receipt of a packet, the node trying to pass the packet to it can try again elsewhere. The route is not specified, only the destination. This is how the internet can hypothetically survive an attack. The nodes keep trying to find neighbors until each packet is eventually routed to its destination.

In practice, the internet as it has evolved is a little less robust than that scenario implies. But the packet architecture is still the core of the design.

The decentralized nature of the architecture makes it almost impossible to track the nature of the information that is flowing through it. Each packet is just a tiny piece of a file, so even if you look at the contents of packets going by, it can sometimes be hard to figure out what the whole file will be when it is reassembled at the destination.

In more recent eras, ideologies related to privacy and anonymity joined a fascination with emerging systems similar to some conceptions of biological evolution to influence engineers to reinforce the opacity of the design of the internet. Each new layer of code has furthered the cause of deliberate obscurity.

Because of the current popularity of cloud architectures, for instance, it has become difficult to know which server you are logging into from time to time when you use particular software. That can be an annoyance in certain circumstances in which latency—the time it takes for bits to travel between computers—matters a great deal.

The appeal of deliberate obscurity is an interesting anthropological question. There are a number of explanations for it that I find to have merit. One is a desire to see the internet come alive as a metaorganism: many engineers hope for this eventuality, and mystifying the workings of the net makes it easier to imagine it is happening. There is also a revolutionary fantasy: engineers sometimes pretend they are assailing a corrupt existing media order and demand both the covering of tracks and anonymity from all involved in order to enhance this fantasy.

At any rate, the result is that we must now measure the internet as if it were a part of nature, instead of from the inside, as if we were examining the books of a financial enterprise. We must explore it as if it were unknown territory, even though we laid it out.

The means of conducting explorations are not comprehensive. Leaving aside ethical and legal concerns, it is possible to "sniff" packets traversing a piece of hardware comprising one node in the net, for instance. But the information available to any one observer is limited to the nodes being observed.

Rage

I well recall the birth of the free software movement, which preceded and inspired the open culture variant. It started out as an act of rage more than a quarter of a century ago.

Visualize, if you will, the most transcendently messy, hirsute, and otherwise eccentric pair of young nerds on the planet. They were in their early twenties. The scene was an uproariously messy hippie apartment in Cambridge, Massachusetts, in the vicinity of MIT. I was one of these men; the other was Richard Stallman.

Stallman was distraught to the point of tears. He had poured his energies into a celebrated project to build a radically new kind of computer called the LISP machine. But it wasn't just a regular computer running LISP, a programming language beloved by artificial intelligence researchers.* Instead, it was a machine patterned on LISP from the bottom up, making a radical statement about what computing could be like at every level, from the underlying architecture to the user interface. For a brief period, every hot computer science department had to own some of these refrigerator-size gadgets.

Eventually a company called Symbolics became the primary seller of LISP machines. Stallman realized that a whole experimental subculture of computer science risked being dragged into the toilet if anything bad happened to a little company like Symbolics— and of course everything bad happened to it in short order.

So Stallman hatched a plan. Never again would computer code, and the culture that grew up with it, be trapped inside a wall of commerce and legality. He would develop a free version of an ascendant, if rather dull, software tool: the UNIX operating system. That simple act would blast apart the idea that lawyers and companies could control software culture.

Eventually a young programmer of the next generation named Linus Torvalds followed in Stallman's footsteps and did something similar, but

Why are so many of the more sophisticated examples of code in the online world— like the page-rank algorithms in the top search engines or like Adobe's Flash—the results of proprietary development? Why did the adored iPhone come out of what many regard as the most closed, tyrannically managed software-development shop on Earth? An honest empiricist must conclude that while the open approach has been able to create lovely, polished copies, it hasn't been so good at creating notable originals. Even though the open-source movement has a stinging countercultural rhetoric, it has in practice been a conservative force.

---

*LISP, conceived in 1958, made programming a computer look approximately like writing mathematical expressions. It was a huge hit in the crossover world between math and computer science starting in the 1960s. Any realization of my proposal for formal financial expression, described in Chapter 7, would undoubtedly bear similarities to LISP.

using the popular Intel chips. In 1991 that effort yielded Linux, the basis for a vastly expanded free software movement.

But back to that dingy bachelor pad near MIT. When Stallman told me his plan, I was intrigued but sad. I thought that code was important in more ways than politics can ever be. If politically motivated code was going to amount to endless replays of relatively dull stuff like UNIX instead of bold projects like the LISP machine, what was the point? Would mere humans have enough energy to sustain both kinds of idealism?

Twenty-five years later, it seems clear that my concerns were justified. Open wisdom-of-crowds software movements have become influential, but they haven't promoted the kind of radical creativity I love most in computer science. If anything, they've been hindrances. Some of the youngest, brightest minds have been trapped in a 1970s intellectual framework because they are hypnotized into accepting old software designs as if they were facts of nature. Linux is a superbly polished copy of an antique—shinier than the original, perhaps, but still defined by it.

I'm not anti–open source. I frequently argue for it in various specific projects. But the politically correct dogma that holds that open source is automatically the best path to creativity and innovation is not borne out by the facts.

A Disappointment Too Big to Notice

How can you know what is lame and derivative in someone else's experience? How can you know if you get it? Maybe there's something amazing happening and you just don't know how to perceive it. This is a tough enough problem when the topic is computer code, but it's even harder when the subject is music.

The whole idea of music criticism is not pleasant to me, since I am, after all, a working musician. There is something confining and demeaning about having expectations of something as numinous as music in the first place. It isn't as if anyone really knows what music is, exactly. Isn't music pure gift? If the magic appears, great, but if it doesn't, what purpose is served by complaining?

But sometimes you have to at least approach critical thinking. Stare into the mystery of music directly, and you might turn into a pillar of salt, but you must at least survey the vicinity to know where not to look.

So it is with the awkward project of assessing musical culture in the age of the internet. I entered the internet era with extremely high expectations. I eagerly anticipated a chance to experience shock and intensity and new sensations, to be thrust into lush aesthetic wildernesses, and to wake up every morning to a world that was richer in every detail because my mind had been energized by unforeseeable art.

Such extravagant expectations might seem unreasonable in retrospect, but that is not how they seemed twenty-five years ago. There was every reason to have high expectations about the art—particularly the music—that would arise from the internet.

Consider the power of music from just a few figures from the last century. Dissonance and strange rhythms produced a riot at the premiere of Stravinsky's *Rite of Spring*. Jazz musicians like Louis Armstrong, James P. Johnson, Charlie Parker, and Thelonius Monk raised the bar for musical intelligence while promoting social justice. A global cultural shift coevolved with the Beatles' recordings. Twentieth-century pop music transformed sexual attitudes on a global basis. Trying to summarize the power of music leaves you breathless.

Changing Circumstances Always Used to Inspire Amazing New Art

It's easy to forget the role technology has played in producing the most powerful waves of musical culture. Stravinsky's *Rite of Spring*, composed in 1912, would have been a lot harder to play, at least at tempo and in tune, on the instruments that had existed some decades earlier. Rock and roll—the electric blues—was to a significant degree a successful experiment in seeing what a small number of musicians could do for a dance hall with the aid of amplification. The Beatles' recordings were in part a rapid reconnaissance mission into the possibilities of multitrack recording, stereo mixes, synthesizers, and audio special effects such as compression and varying playback speed.

Changing economic environments have also stimulated new music in the past. With capitalism came a new kind of musician. No longer tied to the king, the whorehouse, the military parade, the Church, the side-walk busker's cup, or the other ancient and traditional sources of musi-cal patronage, musicians had a chance to diversify, innovate, and be entrepreneurial. For example, George Gershwin made some money from sheet music sales, movie sound tracks, and player piano rolls, as well as from traditional gigs.

So it seemed entirely reasonable to have the highest expectations for music on the internet. We thought there would be an explosion of wealth and of ways to become wealthy, leading to super-Gershwins. A new species of musician would be inspired to suddenly create radically new kinds of music to be performed in virtual worlds, or in the margins of e-books, or to accompany the oiling of fabricating robots. Even if it was not yet clear what business models would take hold, the outcome would surely be more flexible, more open, more hopeful than what had come before in the hobbled economy of physicality.

The Blankness of Generation X Never Went Away, but Became the New Normal

At the time that the web was born, in the early 1990s, a popular trope was that a new generation of teenagers, reared in the conservative Rea-gan years, had turned out exceptionally bland. The members of "Gener-ation X" were characterized as blank and inert. The anthropologist Steve Barnett compared them to pattern exhaustion, a phenomena in which a culture runs out of variations of traditional designs in their pottery and becomes less creative.

A common rationalization in the fledgling world of digital culture back then was that we were entering a transitional lull before a creative storm—or were already in the eye of one. But the sad truth is that we were not passing through a momentary lull before a storm. We had instead entered a persistent somnolence, and I have come to believe that we will only escape it when we kill the hive.

The First-Ever Era of Musical Stasis

Here is a claim I wish I weren't making, and that I would prefer to be wrong about: popular music created in the industrialized world in the

THE UNBEARABLE THINNESS OF FLATNESS

decade from the late 1990s to the late 2000s doesn't have a distinct style—that is, one that would provide an identity for the young people who grew up with it. The process of the reinvention of life through music appears to have stopped.

What once seemed novel—the development and acceptance of un-original pop culture from young people in the mid-1990s (the Gen Xers)—has become so commonplace that we do not even notice it any-more. We've forgotten how fresh pop culture can be.

Where is the new music? Everything is retro, retro, retro.

Music is everywhere, but hidden, as indicated by tiny white prairie dog–like protuberances popping out of everyone's ears. I am used to see-ing people making embarrassingly sexual faces and moaning noises when listening to music on headphones, so it's taken me a while to get used to the stone faces of the earbud listeners in the coffeehouse.

Beating within the retro indie band that wouldn't have sounded out of place even when I was a teenager there might be some exotic heart, some layer of energy I'm not hearing. Of course, I can't know my own limits. I can't know what I am not able to hear.

But I have been trying an experiment. Whenever I'm around "Face-book generation" people and there's music playing—probably selected by an artificial intelligence or crowd-based algorithm, as per the current fashion—I ask them a simple question: Can you tell in what decade the music that is playing right now was made? Even listeners who are not particularly music oriented can do pretty well with this question—but only for certain decades.

Everyone knows that gangster rap didn't exist yet in the 1960s, for instance. And that heavy metal didn't exist in the 1940s. Sure, there's an occasional track that sounds as if it's from an earlier era. Maybe a big-band track recorded in the 1990s might be mistaken for an older record-ing, for instance.

But a decade was always a long time in the development of musical style during the first century of audio recording. A decade gets you from Robert Johnson's primordial blues recordings to Charlie Parker's intensely modernist jazz recordings. A decade gets you from the reign of big bands to the reign of rock and roll. Approximately a decade separated the last Beatles record from the first big-time hip-hop records. In all these examples, it is inconceivable that the later offering could have appeared at the time of the earlier one. I can't find a decade span in the first century of recorded music that didn't involve extreme stylistic evolution, obvious to listeners of all kinds.

We're not just talking about surface features of the music, but the very idea of what music was all about, how it fit into life. Does it convey classiness and confidence, like Frank Sinatra, or help you drop out, like stoner rock? Is it for a dance floor or a dorm room?

There are new styles of music, of course, but they are new only on the basis of technicalities. For instance, there's an elaborate nomenclature for species of similar electronic beat styles (involving all the possible concatenations of terms like dub, house, trance, and so on), and if you learn the details of the nomenclature, you can more or less date and place a track. This is more of a nerd exercise than a musical one—and I realize that in saying that I'm making a judgment that perhaps I don't have a right to make. But does anyone really disagree?

I have frequently gone through a conversational sequence along the following lines: Someone in his early twenties will tell me I don't know what I'm talking about, and then I'll challenge that person to play me some music that is characteristic of the late 2000s as opposed to the late 1990s. I'll ask him to play the tracks for his friends. So far, my theory has held: even true fans don't seem to be able to tell if an indie rock track or a dance mix is from 1998 or 2008, for instance.

I'm obviously not claiming that there has been no new music in the world. And I'm not claiming that all the retro music is disappointing. There are some wonderful musicians in the retro mold, treating old pop music styles as a new kind of classical music and doing so marvelously well.

But I *am* saying that this kind of work is more nostalgic than reaching. Since genuine human experiences are forever unique, pop music of a new era that lacks novelty raises my suspicions that it also lacks authenticity.

There are creative, original musicians at work today, of course. (I hope that on my best days I am one of them.) There are undoubtedly musical marvels hidden around the world. But this is the first time since electrification that mainstream youth culture in the industrialized world has cloaked itself primarily in nostalgic styles.

I am hesitant to share my observations for fear of hexing someone's potentially good online experience. If you are having a great time with music in the online world as it is, don't listen to me. But in terms of the big picture, I fear I am onto something. What of it? Some of my colleagues in the digital revolution argue that we should be more patient; certainly with enough time, culture will reinvent itself. But how patient should we be? I find that I am not willing to ignore a dark age.

Digital Culture That Isn't Retro Is Still Based in a Retro Economy

Even the most seemingly radical online enthusiasts seem to always flock to retro references. The sort of "fresh, radical culture" you expect to see celebrated in the online world these days is a petty mashup of preweb culture.

Take a look at one of the big cultural blogs like Boing Boing, or the endless stream of mashups that appear on YouTube. It's as if culture froze just before it became digitally open, and all we can do now is mine the past like salvagers picking over a garbage dump.

This is embarrassing. The whole point of connected media technologies was that we were supposed to come up with new, amazing cultural expression. No, more than that—we were supposed to invent better fundamental types of expression: not just movies, but interactive virtual worlds; not just games, but simulations with moral and aesthetic profundity. That's why I was criticizing the old way of doing things.

Fortunately, there are people out there engaging in the new kinds of expression that my friends and I longed for at the birth of the web. Will Wright, creator of The Sims and Spore, is certainly creating new-media forms. Spore is an example of the new kind of expression that I had hoped for, the kind of triumph that makes all the hassles of the digital age worthwhile.

The Spore player guides the evolution of simulated alien life-forms. Wright has articu-

Freedom is moot if you waste it. If the internet is really destined to be no more than an ancillary medium, which I would view as a profound defeat, then it at least ought to do whatever it can not to bite the hand that feeds it—that is, it shouldn't starve the commercial media industries.

lated—not in words, but through the creation of a gaming experience—what it would be like to be a god who, while not rethinking every detail of his creation at every moment, occasionally tweaks a self-perpetuating universe.

Spore addresses an ancient conundrum about causality and deities that was far less expressible before the advent of computers. It shows that digital simulation can explore ideas in the form of direct experiences, which was impossible with previous art forms.

Wright offers the hive a way to play with what he has done, but he doesn't create using a hive model. He relies on a large staff of full-time paid people to get his creations shipped. The business model that allows this to happen is the only one that has been proven to work so far: a closed model. You actually pay real money for Wright's stuff.

Wright's work is something new, but his life is of the previous century. The new century is not yet set up to support its own culture. When Spore was introduced, the open culture movement was offended because of the inclusion of digital rights management software, which meant that it wasn't possible for users to make copies without restriction. As punishment for this sin, Spore was hammered by mobs of trolls on Amazon reviews and the like, ruining its public image. The critics also defused what should have been a spectacular debut, since Wright's previous offerings, such as The Sims, had achieved the very pinnacle of success in the gaming world.

Some other examples are the iPhone, the Pixar movies, and all the other beloved successes of digital culture that involve innovation in the result as opposed to the ideology of creation. In each case, these are personal expressions. True, they often involve large groups of collaborators, but there is always a central personal vision—a Will Wright, a Steve Jobs, or a Brad Bird conceiving the vision and directing a team of people earning salaries.

CHAPTER 10

DIGITAL CREATIVITY ELUDES FLAT PLACES

A HYPOTHESIS LINKS the anomaly in popular music to the characteristics of flat information networks that suppress local contexts in favor of global ones.

What Makes Something Real Is That It Is Impossible to Represent It to Completion

It's easy to forget that the very idea of a digital expression involves a trade-off with metaphysical overtones. A physical oil painting cannot convey an image created in another medium; it is impossible to make an oil painting look just like an ink drawing, for instance, or vice versa. But a digital image of sufficient resolution can capture any kind of perceivable image—or at least that's how you'll think of it if you believe in bits too much.

Of course, it isn't really so. A digital image of an oil painting is forever a representation, not a real thing. A real painting is a bottomless mystery, like any other real thing. An oil painting changes with time; cracks appear on its face. It has texture, odor, and a sense of presence and history.

Another way to think about it is to recognize that there is no such thing as a digital object that isn't specialized. Digital representations can be very good, but you can never foresee all the ways a representation might need to be used. For instance, you could define a new MIDIlike standard for representing oil paintings that includes odors, cracks, and so on, but it will always turn out that you forgot something, like the weight or the tautness of the canvas.

The definition of a digital object is based on assumptions of what aspects of it will turn out to be important. It will be a flat, mute nothing if you ask something of it that exceeds those expectations. If you didn't specify the weight of a digital painting in the original definition, it isn't just weightless, it is less than weightless.

A physical object, on the other hand, will be fully rich and fully real whatever you do to it. It will respond to any experiment a scientist can conceive. What makes something fully real is that it is impossible to represent it to completion.

A digital image, or any other kind of digital fragment, is a useful compromise. It captures a certain limited measurement of reality within a standardized system that removes any of the original source's unique qualities. No digital image is really distinct from any other; they can be morphed and mashed up.

That doesn't mean that digital culture is doomed to be anemic. It just means that digital media have to be used with special caution.

Anger in Antisoftware

Computers can take your ideas and throw them back at you in a more rigid form, forcing you to live within that rigidity unless you resist with significant force.

A good example to consider is the humble musical note, which I discussed in the first chapter. People have played musical notes for a very long time. One of the oldest human-hewn extant artifacts is a flute that appears to have been made by Neanderthals about 75,000 years ago. The flute plays approximately in tune. Therefore it is likely that whoever played that old flute had a notion of discrete toots. So the idea of the note goes back very far indeed.

But as I pointed out earlier, no single, precise idea of a note was ever a mandatory part of the process of making music until the early 1980s, when MIDI appeared. Certainly, various ideas about notes were used to notate music before then, as well as to teach and to analyze, but the phenomenon of music was bigger than the concept of a note.

A similar transformation is present in neoclassical architecture. The original classical buildings were tarted up with garish colors and decorations, and their statues were painted to appear more lifelike. But when architects and sculptors attempted to re-create this style long after the paint and ornamentation had faded away, they invented a new cliché: courthouses and statuary made of dull stone.

A neoclassical effect was formalized for music with the invention of MIDI. For the first time, it took effort not to succumb to neoclassical reinvention, even of one's own freshly invented music. This is one of the dangers presented by software tools.

The best music of the web era seems to me to be "antisoftware." The last genuinely new major style was probably hip-hop. That's a rather sad thing to say, since hip-hop has already seen at least three generations of artists. Hip-hop's origins predate the web, as do the origins of every other current style.

But hip-hop has been alive during the web era, or at least not as stuck as the endless repetitions of the pop, rock, and folk genres. The usual narrative one hears within hip-hop culture is that it "appropriated" digital technology—but I hear things differently. Hip-hop is imprisoned within digital tools like the rest of us. But at least it bangs fiercely against the walls of its confinement.

Outside of hip-hop, digital music usually comes off as sterile and bland. Listen to a lot of what comes out of the university computer music world, the world of laptop-generated chill-out music, or new-age ambient music, and you'll hear what I mean. Digital production usually has an overly regular beat because it comes out of a looper or a sequencer. And because it uses samples, you hear identical microstructure in sound again and again, making it seem as if the world is not fully alive while the music is playing.

But hip-hop pierced through this problem in a shocking way. It turns out these same deficits can be turned around and used to express anger with incredible intensity. A sample played again and again expresses stuckness and frustration, as does the regular beat. The inherent rigidity of software becomes a metaphor for an alienated modern life mired in urban poverty. A digital sound sample in angry rap doesn't correspond to the graffiti but to the wall.

Empathy and Locality:
The Blandness of Global Context

The hive ideology robs musicians and other creative people of the ability to influence the context within which their expressions are perceived, if they are to transition out of the old world of labels and music licensing. This is one of the more serious disconnects between what I love about making music and the way it is being transformed by the hive-minded movement. I've gone back and forth endlessly with ideological new-music entrepreneurs who have asked me to place my music into Creative Commons or some other hive scheme.

I have always wanted a simple thing, and the hive refuses to give it to me. I want both to encourage reuse of my music and to interact with the person who hopes to use some of my music in an aggregate work. I might not even demand an ability to veto that other person's plans, but I want at least a chance at a connection.

There are areas of life in which I am ready to ignore the desire for connection in exchange for cash, but if art is the focus, then interaction is what I crave. The whole point of making music for me is connecting with other people. Why should I have to give that up?

But no, that option is not currently supported, and the very notion is frowned upon. Creative Commons, for one, asks you to choose from a rich variety of licensing options. You can demand attribution—or not—when your music is mashed into a compound product, for instance.

Context has always been part of expression, because expression becomes meaningless if the context becomes arbitrary. You could come up with an invented language in which the letters that compose the words to John Lennon's "Imagine" instead spell out the instructions for cleaning a refrigerator. Meaning is only ever meaning in context.

I realize the whole point is to get a lot of free content out there, especially content that can be mashed up, but why won't Creative Commons provide an option along the lines of this: Write to me and tell me what you want to do with my music. If I like it, you can do so immediately. If I don't like what you want to do, you can still do it, but you will have to wait six months. Or, perhaps, you will have to go through six rounds of arguing back and forth with me about it, but then you can do whatever you want. Or you might have to always include a notice in the mashup stating that I didn't like the idea, with my reasons.

Why must all the new schemes that compete with traditional music licensing revere remoteness? There's no significant technological barrier to getting musicians involved in the contextual side of expression, only an ideological one.

The response I usually get is that there's nothing preventing me from collaborating with someone I find by some other means, so what difference does it make if third parties I never know are using the same digital fragments of my music in unrelated ways?

Every artist tries to foresee or even nudge the context in which expression is to be perceived so that the art will make sense. It's not necessarily a matter of overarching ego, or manipulative promotion, but a simple desire for meaning.

A writer like me might choose to publish a book on paper, not only because it is the only way to get decently paid at the moment, but also because the reader then gets the whole book at once, and just might read it as a whole.

When you come upon a video clip or picture or stretch of writing that has been made available in the web 2.0 manner, you almost never have access to the history or the locality in which it was perceived to have meaning by the anonymous person who left it there. A song might have been tender, or brave, or redemptive in context, but those qualities will usually be lost.

Even if a video of a song is seen a million times, it becomes just one dot in a vast pointillist spew of similar songs when it is robbed of its motivating context. Numerical popularity doesn't correlate with intensity of connection in the cloud.

If a fuzzy crowd of anonymous people is making uninformed mashups with my recorded music, then when I present my music myself the context becomes one in which my presentation fits into a statistical distribution of other presentations. It is no longer an expression of my life.

Under those circumstances, it is absurd to think that there is any connection between me and mashers, or those who perceive the mashups. Empathy—connection—is then replaced by hive statistics.

CHAPTER 11

aLL HaIL THe MeMBRane

FLAT GLOBAL NETWORKS are criticized as poor designs for scientific or technical communities. Hierarchical encapsulation is celebrated in natural evolution and human thought.

How Nature Asks Questions

There are some deep principles here that apply far beyond culture and the arts. If you grind any information structure up too finely, you can lose the connections of the parts to their local contexts as experienced by the humans who originated them, rendering the structure itself meaningless. The same mistakes that have stultified some recent digital culture would be disastrous if applied to the sciences, for instance. And yet there is some momentum toward doing just that.

In fact, there is even a tendency to want to think of nature as if she were a hive mind, which she is not. For instance, nature could not maximize the meaning of genes without species.

There's a local system for each species within which creativity is tested. If all life existed in a undifferentiated global gloop, there would be little evolution, because the process of evolution would not be able to ask coherent, differentiated questions.

A Wikified Science Conference

The illusions of the hive mind haven't thus far had as much influence in science as in music, but there's a natural zone of blending of the Silicon Valley and scientific communities, so science hasn't been entirely unaffected.

There are two primary strands of cybernetic totalism. In one strand, the computing cloud is supposed to get smart to a superhuman degree on its own, and in the other, a crowd of people connected to the cloud through anonymous, fragmentary contact is supposed to be the super-human entity that gets smart. In practice the two ideas become similar.

The second, wiki approach has gotten more traction in the scientific community so far. Sci Foo, for instance, is an experimental, invitation-only wikilike annual conference that takes place at Google headquarters in Mountain View, California. There is almost no preplanned agenda. Instead, there's a moment early on when the crowd of scientists rushes up to blank poster-size calendars and scrawls on them to reserve rooms and times for talks on whatever topic comes to mind.

It wasn't official, of course, but the big idea kept popping up at a recent Sci Foo I attended: science as a whole should consider adopting the ideals of web 2.0, becoming more like the community process behind Wikipedia or the open-source operating system Linux. And that goes double for synthetic biology, the current buzzword for a super-ambitious conception of biotechnology that draws on the techniques of computer science. There were more sessions devoted to ideas along these lines than to any other topic, and the presenters of those sessions tended to be the younger people, indicating that the notion is ascendant.

Wikified Biology

There were plenty of calls at Sci Foo for developing synthetic biology along open-source lines. Under such a scheme, DNA sequences might float around from garage experimenter to garage experimenter via the internet, following the trajectories of pirated music downloads and being recombined in endless ways.

The quintessential example of the open ideal showed up in Freeman Dyson's otherwise wonderful piece about the future of synthetic biology in the *New York Review of Books*. MIT bioengineer Drew Endy, one of the enfants terribles of synthetic biology, opened his spectacular talk at Sci Foo with a slide of Dyson's article. I can't express the degree to which I admire Freeman, but in this case, we see things differently.

Dyson equates the beginnings of life on Earth with the Eden of Linux. Back when life first took hold, genes flowed around freely; genetic sequences skipped from organism to organism in much the way they may soon be able to on the internet. In his article, Freeman derides the first organism that hoarded its genes behind a protective membrane as "evil," just like the nemesis of the open-software movement, Bill Gates.

Once organisms became encapsulated, they isolated themselves into distinct species, trading genes only with others of their kind. Freeman suggests that the coming era of synthetic biology will be a return to Eden.

I suppose amateurs, robots, and an aggregation of amateurs and robots might someday hack genes in the global garage and tweet DNA sequences around the globe at light speed. Or there might be a slightly more sober process that takes place between institutions like high schools and start-up companies.

However it happens, species boundaries will become defunct, and genes will fly about, resulting in an orgy of creativity. Untraceable multitudes of new biological organisms will appear as frequently as new videos do on YouTube today.

One common response to suggestions that this might happen is fear. After all, it might take only one doomsday virus produced in one garage to bring the entire human story to a close. I will not focus directly on that concern, but, instead, on whether the proposed style of openness would even bring about the creation of innovative creatures.

The alternative to wide-open development is not necessarily evil. My guess is that a poorly encapsulated communal gloop of organisms lost out to closely guarded species on the primordial Earth for the same reason that the Linux community didn't come up with the iPhone: encapsulation serves a purpose.

Orgies Are Poorly Designed Experiments

Let's say you have something complicated, like a biological cell, or even something much less complicated, like a computer design or a scientific model. You put it through tests, and the results of the tests influence how the design should be changed. That can happen either in natural evolution or in a lab.

The universe won't last long enough for every possible combination of elements in a complicated construction like a cell to be tested. Therefore, the only option is to establish as much as possible from the results of each test and proceed incrementally. After a series of encapsulated tests, it might seem as though an improved result appears magically, as if it couldn't have been approached incrementally.

Fortunately, encapsulation in human affairs doesn't require lawyers or a tyrant; it can be achieved within a wide variety of political structures. Academic efforts are usually well encapsulated, for instance. Scientists don't publish until they are ready, but publish they must. So science as it is already practiced is open, but in a punctuated, not continuous, way. The interval of nonopenness—the time before publication—functions like the walls of a cell. It allows a complicated stream of elements to be defined well enough to be explored, tested, and then improved.

The politically incorrect critique of Freeman's point of view is that the restrictions created by species boundaries have similarly made billions of years of natural biology more like hardware than like software. Hardware is the stuff that improves according to that exponential demon, Moore's law, because there's a box around it and you can tell what it's doing. Software is the stuff that rarely, if ever, improves. There is no box around it, no way to predict all the interactions it might have to endure.

To put it another way: there won't be an orgy of creativity in an overly open version of synthetic biology, because there have to be species for sex to make sense.

> The open-source software community is simply too connected to focus its tests and maintain its criteria over an extended duration. A global process is no test at all, for the world happens only once. You need locality to have focus, evolution, or any other creative process.

You Don't Know What You're Missing

If Linux provides one model for the future of open culture and science, Wikipedia provides another.

Many scientists, especially younger ones, hold Wikipedia in high regard. I don't dispute many of the achievements claimed by proponents of Wikipedia. The problems I worry about are perhaps subtle, but I think they are important nonetheless.

Wikipedia is a great example of the dilemma I face when I argue,

"You don't know what you're missing." The collective encyclopedia is used by almost everyone at this point, so what's the problem?

There seems to be no limit to Wikipedia adoration. For example, a ghastly news story—such as one covering a terrorist event—might focus on how magically the corresponding Wikipedia entry came together, as if that were the situation's silver lining.*

I am not strictly against any particular digital technology. There is nothing wrong with using Wikipedia—in moderation. I do myself. But I'd like to engage the reader in challenging the elevated position Wikipedia has been granted in the online environment.

As a source of useful information, Wikipedia excels in two areas: pop culture and hard science. In the first category, truth is fiction anyway, so what the wiki says is by definition true; in the second, there actually is a preferred truth, so it is more plausible to speak with a shared voice.

Wikipedia was predicted by Douglas Adams's science fiction comedy *Hitchhiker's Guide to the Galaxy*. His fictional *Guide* functioned in a similar way, with one of its contributors able to instantaneously update the entire entry for Planet Earth (from "Harmless" to "Mostly harmless") with a few taps on a keyboard. Though Earth merited a two-word entry, there were substantial articles about other topics, such as which alien poetry was the worst and how to make strange cocktails. The first thought is often the best thought, and Adams perfectly captured the spirit of much of Wikipedia before it was born.

It has been pointed out that Wikipedia entries about geeky pop culture are longer and more lovingly crafted than those regarding reality. A science fiction army from a movie or novel will typically be better described than an army from reality; a porn star will get a more detailed biography than a Nobel Prize winner.†

This is not the aspect of Wikipedia that I dislike. It's great that we now enjoy a cooperative pop culture concordance. This is where the Wikipedians take on true voices: they become human when they reveal

---

*See Norm Cohen, "The Latest on Virginia Tech, from Wikipedia," *New York Times*, April 23, 2007. In 2009, Twitter became the focus of similar stories because of its use by protestors of Iran's disputed presidential election.

†See Jamin Brophy-Warren, "Oh, That John Locke," *Wall Street Journal*, June 16, 2007.

themselves. However, one is constantly bombarded with declarations about how amazingly useful and powerful Wikipedia is with regard to nonfiction topics. These are not untrue statements, but they can be misleading.

If you want to see how valuable something is, try living without it for a while. Spend some time ignoring Wikipedia. When you look something up in a search engine, just keep flipping through results until you find the first one written by a particular person with a connection to the topic. If you do this, you'll generally find that for most topics, the Wikipedia entry is the first URL returned by search engines but not necessarily the best URL available.

It seems to me that if Wikipedia suddenly disappeared, similar information would still be available for the most part, but in more contextualized forms, with more visibility for the authors and with a greater sense of style and presence—though some might counter that the non-Wikipedia information is not organized in as consistent and convenient a way.

The convenience factor is real, but part of the reason is that Wikipedia provides search engines with a way to be lazy. There really is no longer any technology behind the choice of the first result for a great many searches. Especially on mobile devices, text-entry boxes and software widgets that are devoted purely to Wikipedia are starting to appear, not even bothering to include the web at large. If Wikipedia is treated as the overarching, primary text of the human experience, then of course it will, as if by decree, become "more convenient" than other texts.

Another part of the convenience factor is the standardization of presentation. While I've run across quite a few incomprehensible, terribly written passages in Wikipedia articles, on the whole there's a consistency of style. This can be either a benefit or a loss, depending on the topic and what you are after. Some topics need the human touch and a sense of context and personal voice more than others.

Do Edit Wars Have Casualties?

One of the negative aspects of Wikipedia is this: because of how its entries are created, the process can result in a softening of ambition or, more specifically, a substitution of ideology for achievement.

Discussions of Wikipedia usually center on the experience of people who use it as a resource. That's important, but I would like to also focus on the experience of the people who create it. They aren't a random assortment of people, even if they sometimes pretend to be. They are often, so far as I can tell, people who are committed to whatever area they are writing about.

Science-related Wikipedia entries often come together in a cordial manner because the scientific community is practiced at being cordial. So the experience of scientists writing in Wikipedia is probably better on average than it is for other contributors.

Typical authors of Wikipedia, however, implicitly celebrate the ideal of intellectual mob rule. "Edit wars" on Wikipedia are called that for a reason. Whether they are cordial or not, Wikipedians always act out the idea that the collective is closer to the truth and the individual voice is dispensable.

To understand the problem, let's focus on hard science, the area aside from pop culture where Wikipedia seems to be the most reliable. In fact, let's consider the hardest of the hard: math.

Math as Expression

For many people math is hard to learn, and yet to those who love it, doing math is a great joy that goes beyond its obvious utility and puts it in an aesthetic realm. Albert Einstein called it "the poetry of logical ideas."

Math is an arena in which it's appropriate to have high hopes for the future of digital media. A superb development—which might take place in decades or centuries to come—would be for some new channel of communication to come along that makes a deep appreciation of math more widely available. Then the fundamental patterning of reality, which only math can describe, would become part of a wider human conversation.

This kind of development might follow the course of what has happened to moviemaking. It used to be that movies came only from a few elite studios that had access to the expensive and cumbersome equipment then necessary to make films. Now anyone can make a movie; moviemaking has become a part of general experience.

The reason moviemaking has become as much a part of pop culture as movie viewing is that new gadgets appeared. Cheap, easy-to-use video cameras, editing software, and distribution methods—such as YouTube—are what made the difference. Before them, it might have seemed as though moviemaking was such an esoteric practice that even if widely accessible tools arrived, the experience would still only be available to a few special geniuses.

And while it's true that there are still only a few special geniuses of cinema, the basic competence turns out to be as easily acquired as learning to talk or drive a car. The same thing ought to happen to math someday. The right tools could help math become another way large numbers of people can connect creatively in our culture.

In the late 1990s I was excited because it looked as if it was starting to happen. All over the world, mathematicians of all stripes were beginning to create websites that explored the potential for explaining what they do for civilians. There were online introductions to wonderful geometric shapes, strange knots of logic, and magical series of numbers. None of the material was perfect; in fact, most of it was strange and awkward. But this kind of mass development was something that had never happened before on such a scale and with such a variety of participants, so every little detail was an experiment. It was slow going, but there was a trend that might have led somewhere.

A Forgotten Alternative to Wikis

One institution from this nearly forgotten chapter of the early web was ThinkQuest. This was a contest run by internet pioneers, especially Al Weis, in which teams of high school students competed for scholarships by designing websites that explained ideas from a wide variety of academic disciplines, including math.

Early on, ThinkQuest enjoyed a successful niche similar to the one Wikipedia occupies today. A nonprofit site, it was drawing the same huge numbers of visitors as the big commercial sites of the era, which included some outfits with names like AOL. A ThinkQuest entry was often the first result of a web search.

But the contributions of ThinkQuest were far more original and valuable than those of Wikipedia. The contestants had to learn how to present ideas as wholes, as well as figure out how to use the new online medium to do that. Their work included simulations, interactive games, and other elements that were pretty new to the world. They weren't just transferring material that already existed into a more regularized, anonymous form.

ThinkQuest probably cost a little more than Wikipedia to operate because the machinery of judging used experts—it wasn't supposed to be a war or a popularity contest—but it was still cheap.

The search for new ways to share math on the web was and is incredibly hard work.* Most ThinkQuest entries were poor, and the ones that were good required extraordinary effort.

The web should have developed along the ThinkQuest model instead of the wiki model—and would have, were it not for hive ideology.

When Search Was Hogged

For a few years, there were often multiple pages of top results to a great many queries in search engines like Google that were really just echoes of a Wikipedia entry. It was as if Wikipedia were the only searchable web page for a big slice of human thought and experience. The situation seems to have become better recently—I assume because search engines have responded to complaints.

People who contribute to Wikipedia naturally become emotionally committed to what they have done. Their vain links probably helped drive the search engines to the one book of the hive. But the era when

---

*For example, figuring out how to present a hendecachoron, which is a four-dimensional shape I love, in an accessible, interactive web animation is an incredibly hard task that has still not been completed. By contrast, contributing to a minimal, raw, dry, but accurate entry about a hendecachoron on Wikipedia is a lot easier, but it offers nothing to someone encountering the shape for the first time.

This shape is amazing because it is symmetrical like a cube, which has six faces, but the symmetry is of a prime number, eleven, instead of a divisible number like six. This is weird, because prime numbers can't be broken into sets of identical parts, so it sounds a little odd that there could be prime-numbered geometric symmetries. It's possible only because the hendecachoron doesn't fit inside a sphere, in the way a cube can. It fits, instead, along the contours of a close cousin of the sphere, which is called the real projective plane. This shape is like a doubly extreme version of the famous Klein bottle. None other than Freeman Dyson made me aware of the hendecachoron, and Carlo Sequin and I worked on producing the first-ever image of one.

search was hogged made the genuinely creative, struggling, experimental web designs become less visible and less appreciated, often leading to a death spiral.

Much of the older, more personal, and more ambitious material from the first wave of web expression is still out there. If you search online for math and ignore the first results, which are often the Wikipedia entry and its echoes, you start to come across weird individual efforts and even some old ThinkQuest pages. They were often last updated around the time Wikipedia arrived. Wikipedia took the wind out of the trend.*

The quest to bring math into the culture continues, but mostly not online. A huge recent step was the publication of a book on paper by John Conway, Heidi Burgiel, and Chaim Goodman-Strauss called *The Symmetries of Things*. This is a tour de force that fuses introductory material with cutting-edge ideas by using a brash new visual style. It is disappointing to me that pioneering work continues primarily on paper, having become muted online.

The same could be said about a great many topics other than math. If you're interested in the history of a rare musical instrument, for instance, you can delve into the internet archive and find personal sites devoted to it, though they probably were last updated around the time Wikipedia came into being. Choose a topic you know something about and take a look.

Wikipedia has already been elevated into what might be a permanent niche. It might become stuck as a fixture, like MIDI or the Google ad exchange services. That makes it important to be aware of what you might be missing. Even in a case in which there is an objective truth that is already known, such as a mathematical proof, Wikipedia distracts the potential for learning how to bring it into the conversation in new ways. Individual voice—the opposite of wikiness—might not matter to mathematical truth, but it is the core of mathematical communication.

*Once again, I have to point out that where Wikipedia is useful, it might not be *uniquely* useful. For instance, there is an alternative choice for a site with raw, dry math definitions, run as a free service by a company that makes software for mathematicians. Go to http://mathworld.wolfram.com/.

PART FOUR

MAKING THE BEST
OF BITS

IN THIS SECTION, I will switch to a more positive perspective, examining what distinguishes cybernetic totalism from humanism by considering the evolution of human culture.

What I hope to demonstrate is that each way of thinking has its proper place and a specific, pragmatic scope, within which it makes sense.

We should reject cybernetic totalism as a basis for making most decisions but recognize that some of its ideas can be useful methods of understanding.

The distinction between understanding and creed, between science and ethics, is subtle. I can hardly claim to have mastered it, but I hope the following reports of my progress will be of use.

CHAPTER 12

I am a Contrarian Loop

VARIETIES OF COMPUTATIONALISM are distinguished; realistic computationalism is defined.

The Culture of Computationalism

In Silicon Valley you will meet Buddhists, anarchists, goddess worshippers, Ayn Rand fanatics, self-described Jesus freaks, nihilists, and plenty of libertarians, as well as surprising blends of all of the above and many others who seem to be nonideological. And yet there is one belief system that doesn't quite mesh with any of these identities that nonetheless serves as a common framework.

For lack of a better word, I call it computationalism. This term is usually used more narrowly to describe a philosophy of mind, but I'll extend it to include something like a culture. A first pass at a summary of the underlying philosophy is that the world can be understood as a computational process, with people as subprocesses.

In this chapter I will explore the uses of computationalism in scientific speculation. I will argue that even if you find computationalism helpful in understanding science, it should not be used in evaluating certain kinds of engineering.

Three Less-Than-Satisfying Flavors of Computationalism

Since I'm a rarity in computer science circles—a computationalism critic—I must make clear that computationalism has its uses.

Computationalism isn't always crazy. Sometimes it is embraced because avoiding it can bring about other problems. If you want to consider people as special, as I have advised, then you need to be able to say at least a little bit about where the specialness begins and ends. This is similar to, or maybe even coincident with, the problem of positioning the circle of empathy, which I described in Chapter 2. If you hope for technology to be designed to serve people, you must have at least a rough idea of what a person is and is not.

But there are cases in which any possible setting of a circle can cause problems. Dividing the world into two parts, one of which is ordinary—deterministic or mechanistic, perhaps—and one of which is mystifying, or more abstract, is particularly difficult for scientists. This is the dreaded path of dualism.

It is awkward to study neuroscience, for instance, if you assume that the brain is linked to some other entity—a soul—on a spirit plane. You have to treat the brain simply as a mechanism you don't understand if you are to improve your understanding of it through experiment. You can't declare in advance what you will and will not be able to explain.

I am contradicting myself here, but the reason is that I find myself playing different roles at different times. Sometimes I am designing tools for people to use, while at other times I am working with scientists trying to understand how the brain works.

Perhaps it would be better if I could find one single philosophy that I could apply equally to each circumstance, but I find that the best path is to believe different things about aspects of reality when I play these different roles or perform different duties.

Up to this point, I have described what I believe when I am a technologist. In those instances, I take a mystical view of human beings. My first priority must be to avoid reducing people to mere devices. The best way to do that is to believe that the gadgets I can provide are inert tools and are only useful because people have the magical ability to communicate meaning through them.

When I put on a different hat—that of a collaborator with scientists—then I believe something else. In those cases, I prefer ideas that don't involve magical objects, for scientists can study people as if we were not magical at all. Ideally, a scientist ought to be able to study something a bit without destroying it. The whole point of technology, though, is to change the human situation, so it is absurd for humans to aspire to be inconsequential.

In a scientific role, I don't recoil from the idea that the brain is a kind of computer, but there is more than one way to use computation as a source of models for human beings. I'll discuss three common flavors of computationalism and then describe a fourth flavor, the one that I prefer. Each flavor can be distinguished by a different idea about what would be needed to make software as we generally know it become more like a person.

One flavor is based on the idea that a sufficiently voluminous computation will take on the qualities we associate with people—such as, perhaps, consciousness. One might claim Moore's law is inexorably leading to superbrains, superbeings, and, perhaps, ultimately, some kind of global or even cosmic consciousness. If this language sounds extreme, be aware that this is the sort of rhetoric you can find in the world of Singularity enthusiasts and extropians.

If we leave aside the romance of this idea, the core of it is that meaning arises in bits as a result of magnitude. A set of one thousand records in a database that refer to one another in patterns would not be meaningful without a person to interpret it; but perhaps a quadrillion or a googol of database entries can mean something in their own right, even if there is no being explaining them.

Another way to put it is that if you have enough data and a big and fast enough computer, you can conceivably overcome the problems associated with logical positivism. Logical positivism is the idea that a sentence or another fragment—something you can put in a computer file— means something in a freestanding way that doesn't require invoking the subjectivity of a human reader. Or, to put it in nerd-speak: "The meaning of a sentence is the instructions to verify it."

Logical positivism went out of fashion, and few would claim its banner these days, but it's enjoying an unofficial resurgence with a computer assist. The new version of the idea is that if you have a lot of data, you can make logical positivism work on a large-scale statistical basis. The thinking goes that within the cloud there will be no need for the numinous halves of traditional oppositions such as syntax/semantics, quantity/quality, content/context, and knowledge/wisdom.

A second flavor of computationalism holds that a computer program with specific design features—usually related to self-representation and circular references—is similar to a person. Some of the figures associated with this approach are Daniel Dennett and Douglas Hofstadter, though each has his own ideas about what the special features should be.

Hofstadter suggests that software that includes a "strange loop" bears a resemblance to consciousness. In a strange loop, things are nested within things in such a way that an inner thing is the same as an outer thing.

If you descend on a city using a parachute, land on a roof, enter the building through a door on that roof, go into a room, open another door to a closet, enter it, and find that there is no floor in the closet and you are suddenly once again falling in the vast sky toward the city, you are in a strange loop. The same notion can perhaps be applied to mental phenomena, when thoughts within thoughts lead to the original thoughts. Perhaps that process has something to do with self-awareness—and what it is to be a person.

A third flavor of computationalism is found in web 2.0 circles. In this case, any information structure that *can* be perceived by some real human to also be a person *is* a person. This idea is essentially a revival of the Turing test. If you can perceive the hive mind to be recommending music to you, for instance, then the hive is effectively a person.

I have to admit that I don't find any of these three flavors of computationalism to be useful on those occasions when I put on my scientist's hat.

The first idea, that quantity equals quality in software, is particularly galling, since a computer scientist spends much of his time struggling with the awfulness of what happens to software—as we currently know how to make it, anyway—when it gets large.

The second flavor is also not helpful. It is fascinating and clever to create software with self-representations and weird loopy structures. Indeed, I have implemented the skydiving scenario in a virtual world. I have never observed any profound change in the capabilities of software systems based on an enhanced degree of this kind of trickery, even though there is still a substantial community of artificial intelligence researchers who expect that benefit to appear someday.

As for the third flavor—the pop version of the Turing test—my complaint ought to be clear by now. People can make themselves believe in all sorts of fictitious beings, but when those beings are perceived as inhabiting the software tools through which we live our lives, we have to change ourselves in unfortunate ways in order to support our fantasies. We make ourselves dull.

But there are more ways than these three to think about people as being special from a computational point of view.

Realistic Computationalism

The approach to thinking about people computationally that I prefer, on those occasions when such thinking seems appropriate to me, is what I'll call "realism." The idea is that humans, considered as information systems, weren't designed yesterday, and are not the abstract playthings of some higher being, such as a web 2.0 programmer in the sky or a cosmic Spore player. Instead, I believe humans are the result of billions of years of implicit, evolutionary study in the school of hard knocks. The cybernetic structure of a person has been refined by a very large, very long, and very deep encounter with physical reality.

From this point of view, what can make bits have meaning is that their patterns have been hewn out of so many encounters with reality that they aren't really abstractable bits anymore, but are instead a nonabstract continuation of reality.

Realism is based on specifics, but we don't yet know—and might never know—the specifics of personhood from a computational point of view. The best we can do right now is engage in the kind of storytelling that evolutionary biologists sometimes indulge in.

Eventually data and insight might make the story more specific, but for the moment we can at least construct a plausible story of ourselves in terms of grand-scale computational natural history. A myth, a creation tale, can stand in for a while, to give us a way to think computationally that isn't as vulnerable to the confusion brought about by our ideas about ideal computers (i.e., ones that only have to run small computer programs).

Such an act of storytelling is a speculation, but a speculation with a purpose. A nice benefit of this approach is that specifics tend to be more colorful than generalities, so instead of algorithms and hypothetical abstract computers, we will be considering songbirds, morphing cephalopods, and Shakespearean metaphors.

CHAPTER 13

One STORY OF HOW Semantics MIGHT Have evolved

THIS CHAPTER PRESENTS a pragmatic alternation between philosophies (instead of a demand that a single philosophy be applied in all seasons). Computationalism is applied to a naturalistic speculation about the origins of semantics.

Computers Are Finally Starting to Be Able to Recognize Patterns

In January 2002 I was asked to give an opening talk and performance for the National Association of Music Merchants,[*] the annual trade show for makers and sellers of musical instruments. What I did was create a rhythmic beat by making the most extreme funny faces I could in quick succession.

A computer was watching my face through a digital camera and generating varied opprobrious percussive sounds according to which funny face it recognized in each moment.[†] (Keeping a rhythm with your face is

---

[*]Given my fetish for musical instruments, the NAMM is one of the most dangerous—i.e., expensive—events for me to attend. I have learned to avoid it in the way a recovering gambler ought to avoid casinos.

[†]The software I used for this was developed by a small company called Eyematic, where I served for a while as chief scientist. Eyematic has since folded, but Hartmut Neven and many of the original students started a successor company to salvage the software. That company was swallowed up by Google, but what Google plans to do with the stuff isn't clear yet. I hope they'll come up with some creative applications along with the expected searching of images on the net.

a strange new trick—we should expect a generation of kids to adopt the practice en masse any year now.)

This is the sort of deceptively silly event that should be taken seriously as an indicator of technological change. In the coming years, pattern-recognition tasks like facial tracking will become commonplace. On one level, this means we will have to rethink public policy related to privacy, since hypothetically a network of security cameras could automatically determine where everyone is and what faces they are making, but there are many other extraordinary possibilities. Imagine that your avatar in Second Life (or, better yet, in fully realized, immersive virtual reality) was conveying the subtleties of your facial expressions at every moment.

But until recently, computers couldn't even see a smile. Facial expressions were imbedded deep within the imprecise domain of quality, not anywhere close to the other side, the infinitely deciphered domain of quantity. No smile was precisely the same as any other, and there was no way to say precisely what all the smiles had in common. Similarity was a subjective perception of interest to poets— and irrelevant to software engineers.

There's an even deeper significance to facial tracking. For many years there was an absolute, unchanging divide between what you could and could not represent or recognize with a computer. You could represent a precise quantity, such as a number, but you could not represent an approximate holistic quality, such as an expression on a face.

While there are still a great many qualities in our experience that cannot be represented in software using any known technique, engineers have finally gained the ability to create software that can represent a smile, and write code that captures at least part of what all smiles have in common. This is an unheralded transformation in our abilities that took place around the turn of our new century. I wasn't sure I would live to see it, though it continues to surprise me that engineers and scientists I run across from time to time don't realize it has happened.

Pattern-recognition technology and neuroscience are growing up together. The software I used at NAMM was a perfect example of this intertwining. Neuroscience can inspire practical technology rather quickly. The original project was undertaken in the 1990s under the auspices of Christoph von der Malsburg, a University of Southern California neuroscientist, and his students, especially Hartmut Neven. (Von der Malsburg might be best known for his crucial observation in the early 1980s that synchronous firing—that is, when multiple neurons go off at the same moment—is important to the way that neural networks function.)

In this case, he was trying to develop hypotheses about what functions are performed by particular patches of tissue in the visual cortex—the part of the brain that initially receives input from the optic nerves. There aren't yet any instruments that can measure what a large, complicated neural net is doing in detail, especially while it is part of a living brain, so scientists have to find indirect ways of testing their ideas about what's going on in there.

One way is to build the idea into software and see if it works. If a hypothesis about what a part of the brain is doing turns out to inspire a working technology, the hypothesis certainly gets a boost. But it isn't clear how strong a boost. Computational neuroscience takes place on an imprecise edge of scientific method. For example, while facial expression tracking software might seem to reduce the degree of ambiguity present in the human adventure, it actually might add more ambiguity than it takes away. This is because, strangely, it draws scientists and engineers into collaborations in which science gradually adopts methods that look a little like poetry and storytelling. The rules are a little fuzzy, and probably will remain so until there is vastly better data about what neurons are actually doing in a living brain.

For the first time, we can at least tell the outlines of a reasonable story about how your brain is recognizing things out in the world—such as smiles—even if we aren't sure of how to tell if the story is true. Here is that story . . .

What the World Looks Like to a Statistical Algorithm

I'll start with a childhood memory. When I was a boy growing up in the desert of southern New Mexico, I began to notice patterns on the dirt roads created by the tires of passing cars. The roads had wavy corduroy-like rows that were a little like a naturally emerging, endless sequence of speed bumps. Their spacing was determined by the average speed of the drivers on the road.

When your speed matched that average, the ride would feel less bumpy. You couldn't see the bumps with your eyes except right at sunset, when the horizontal red light rays highlighted every irregularity in the ground. At midday you had to drive carefully to avoid the hidden information in the road.

Digital algorithms must approach pattern recognition in a similarly indirect way, and they often have to make use of a common procedure that's a little like running virtual tires over virtual bumps. It's called the Fourier transform. A Fourier transform detects how much action there is at particular "speeds" (frequencies) in a block of digital information.

Think of the graphic equalizer display found on audio players, which shows the intensity of the music in different frequency bands. The Fourier transform is what does the work to separate the frequency bands.)

Unfortunately, the Fourier transform isn't powerful enough to recognize a face, but there is a related but more sophisticated transform, the Gabor wavelet transform, that can get us halfway there. This mathematical process identifies individual blips of action at particular frequencies in particular places, while the Fourier transform just tells you what frequencies are present overall.

There are striking parallels between what works in engineering and what is observed in human brains, including a Platonic/Darwinian duality: a newborn infant can track a simple diagrammatic face, but a child needs to see people in order to learn how to recognize individuals.

I'm happy to report that Hartmut's group earned some top scores in a government-sponsored competition in facial recognition. The National Institute of Standards and Technology tests facial recognition systems in the same spirit in which drugs and cars are tested: the public needs to know which ones are trustworthy.

From Images to Odors

So now we are starting to have theories—or at least are able to tell detailed stories—about how a brain might be able to recognize features of its world, such as a smile. But mouths do more than smile. Is there a way to extend our story to explain what a word is, and how a brain can know a word?

It turns out that the best way to consider that question might be to consider a completely different sensory domain. Instead of sights or sounds, we might best start by considering the odors detected by a human nose.

For twenty years or so I gave a lecture introducing the fundamentals of virtual reality. I'd review the basics of vision and hearing as well as of touch and taste. At the end, the questions would begin, and one of the first ones was usually about smell: Will we have smells in virtual reality machines anytime soon?

Maybe, but probably just a few. Odors are fundamentally different from images or sounds. The latter can be broken down into primary components that are relatively straightforward for computers—and the brain—to process. The visible colors are merely words for different wavelengths of light. Every sound wave is actually composed of numerous sine waves, each of which can be easily described mathematically. Each one is like a particular size of bump in the corduroy roads of my childhood.

In other words, both colors and sounds can be described with just a few numbers; a wide spectrum of colors and tones is described by the interpolations between those numbers. The human retina need be sensitive to only a few wavelengths, or colors, in order for our brains to process all the intermediate ones. Computer graphics work similarly: a screen of pixels, each capable of reproducing red, green, or blue, can produce approximately all the colors that the human eye can see.[*] A music synthesizer can be thought of as generating a lot of sine waves, then layering them to create an array of sounds.

Odors are completely different, as is the brain's method of sensing them. Deep in the nasal passage, shrouded by a mucous membrane, sits a patch of tissue—the olfactory epithelium—studded with neurons that detect chemicals. Each of these neurons has cup-shaped proteins called olfactory receptors. When a particular molecule happens to fall into a

[*]Current commercial displays are not quite aligned with human perception, so they can't show all the colors we can see, but it is possible that future displays will show the complete gamut perceivable by humans.

matching receptor, a neural signal is triggered that is transmitted to the brain as an odor. A molecule too large to fit into one of the receptors has no odor. The number of distinct odors is limited only by the number of olfactory receptors capable of interacting with them. Linda Buck of the Fred Hutchinson Cancer Research Center and Richard Axel of Columbia University, winners of the 2004 Nobel Prize in Physiology or Medicine, have found that the human nose contains about one thousand different types of olfactory neurons, each type able to detect a particular set of chemicals.

This adds up to a profound difference in the underlying structure of the senses—a difference that gives rise to compelling questions about the way we think, and perhaps even about the origins of language. There is no way to interpolate between two smell molecules. True, odors can be mixed together to form millions of scents. But the world's smells can't be broken down into just a few numbers on a gradient; there is no "smell pixel." Think of it this way: colors and sounds can be measured with rulers, but odors must be looked up in a dictionary.

That's a shame, from the point of view of a virtual reality technologist. There are thousands of fundamental odors, far more than the handful of primary colors. Perhaps someday we will be able to wire up a person's brain in order to create the illusion of smell. But it would take a lot of wires to address all those entries in the mental smell dictionary. Then again, the brain must have some way of organizing all those odors. Maybe at some level smells do fit into a pattern. Maybe there's a smell pixel after all.

Were Odors the First Words?

I've long discussed this question with Jim Bower, a computational neuroscientist at the University of Texas at San Antonio, best known for making biologically accurate computer models of the brain. For some years now, Jim and his laboratory team have been working to understand the brain's "smell dictionary."

They suspect that the olfactory system is organized in a way that has little to do with how an organic chemist organizes molecules (for instance, by the number of carbon atoms on each molecule). Instead, it more closely resembles the complex way that chemicals are associated in the real world. For example, a lot of smelly chemicals—the chemicals that trigger olfactory neurons—are tied to the many stages of rotting or ripening of organic materials. As it turns out, there are three major, distinct chemical paths of rotting, each of which appears to define a different stream of entries in the brain's dictionary of smells.

To solve the problem of olfaction—that is, to make the complex world of smells quickly identifiable—brains had to have evolved a specific type of neural circuitry, Jim believes. That circuitry, he hypothesizes, formed the basis for the cerebral cortex—the largest part of our brain, and per-haps the most critical in shaping the way we think. For this reason, Jim has proposed that the way we think is fundamentally based in the olfactory.

Keep in mind that smells are not patterns of energy, like images or sounds. To smell an apple, you physically bring hundreds or thousands of apple molecules into your body. You don't smell the entire form; you steal a piece of it and look it up in your smell dictionary for the larger reference.

A smell is a synecdoche: a part standing in for the whole. Consequently, smell requires additional input from the other senses. Context is everything: if you are blindfolded in a bathroom and a good French cheese is placed under your nose, your interpretation of the odor will likely be very different than it would be if you knew you were standing in a kitchen. Similarly, if you can see the cheese, you can be fairly confident that what you're smelling is cheese, even if you're in a restroom.

Recently, Jim and his students have been looking at the olfactory systems of different types of animals for evidence that the cerebral cortex as a whole grew out of the olfactory system. He often refers to the olfactory parts of the brain as the "Old Factory," as they are remarkably similar across species, which suggests that the structure has ancient origins. Because smell recognition often requires input from other senses, Jim is particularly interested to know how that input makes its way into the olfactory system.

In fish and amphibians (the earliest vertebrates), the olfactory system sits right next to multimodal areas of the cerebral cortex, where the processing of the different senses overlaps. The same is true in reptiles, but in addition, their cortex has new regions in which the senses are separated. In mammals, incoming sights, sounds, and sensations undergo many processing steps before ending up in the region of overlap. Think of olfaction as a city center and the other sensory systems as sprawling suburbs, which grew as the brain evolved and eventually became larger than the old downtown.

All of which has led Jim and me to wonder: Is there a relationship between olfaction and language, that famous product of the human cerebral cortex? Maybe the dictionary analogy has a real physical basis.

Olfaction, like language, is built up from entries in a catalog, not from infinitely morphable patterns. Moreover, the grammar of language is primarily a way of fitting those dictionary words into a larger context. Perhaps the grammar of language is rooted in the grammar of smell. Perhaps the way we use words reflects the deep structure of the way our brain processes chemical information. Jim and I plan to test this hypothesis by studying the mathematical properties that emerge during computer simulations of the neurology of olfaction.

If that research pans out, it might shed light on some other connections we've noticed. As it happens, the olfactory system actually has two parts: one detects general odors, and the other, the *pheremonic* system, detects very specific, strong odors given off by other animals (usually of the same species), typically related to fear or mating. But the science of olfaction is far from settled, and there's intense controversy about the importance of pheromones in humans.

Language offers an interesting parallel. In addition to the normal language we all use to describe objects and activities, we reserve a special language to express extreme emotion or displeasure, to warn others to watch out or get attention. This language is called swearing.

There are specific neural pathways associated with this type of speech; some Tourette's patients, for instance, are known to swear uncontrollably. And it's hard to overlook the many swear words that are related to orifices or activities that also emit pheremonic olfactory signals. Could there be a deeper connection between these two channels of "obscenity"?

Clouds Are Starting to Translate

Lngwidge iz a straynge thingee. You can probably read that sentence without much trouble. Sentence also not this time hard.

You can screw around quite a bit with both spelling and word order and still be understood. This shouldn't be surprising: language is flexible enough to evolve into new slang, dialects, and entirely new tongues.

In the 1960s, many early computer scientists postulated that human language was a type of code that could be written down in a neat, compact way, so there was a race to crack that code. If it could be deciphered, then a computer ought to be able to speak with people! That end result turned out to be extremely difficult to achieve. Automatic language translation, for instance, never really took off.

In the first decade of the twenty-first century, computers have gotten so powerful that it has become possible to shift methods. A program can look for correlations in large amounts of text. Even if it isn't possible to capture all the language variations that might appear in the real world (such as the above oddities I used as examples), a sufficiently huge number of correlations eventually yields results.

For instance, suppose you have a lot of text in two languages, such as Chinese and English. If you start searching for sequences of letters or characters that appear in each text under similar circumstances, you can start to build a dictionary of correlations. That can produce significant results, even if the correlations don't always fit perfectly into a rigid organizing principle, such as a grammar.

Such brute-force approaches to language translation have been demonstrated by companies like Meaningful Machines, where I was an adviser for a while, and more recently by Google and others. They can be incredibly inefficient, often involving ten thousand times as much computation as older methods—but we have big enough computers in the clouds these days, so why not put them to work?

Set loose on the internet, such a project could begin to erase language barriers. Even though automatic language translation is unlikely to become as good as what a human translator can do anytime soon, it might get good enough—perhaps not too far in the future—to make countries and cultures more transparent to one another.

Editing Is Sexy; Creativity Is Natural

These experiments in linguistic variety could also inspire a better understanding of how language came about in the first place. One of Charles

Darwin's most compelling evolutionary speculations was that music might have preceded language. He was intrigued by the fact that many species use song for sexual display and wondered if human vocalizations might have started out that way too. It might follow, then, that vocalizations could have become varied and complex only later, perhaps when song came to represent actions beyond mating and such basics of survival.

Language might not have entirely escaped its origins. Since you can be understood even when you are not well-spoken, what is the point of being well-spoken at all? Perhaps speaking well is still, in part, a form of sexual display. By being well-spoken I show not only that I am an intelligent, clued-in member of the tribe but also that I am likely to be a successful partner and helpful mate.

Only a handful of species, including humans and certain birds, can make a huge and ever-changing variety of sounds. Most animals, including our great-ape relatives, tend to repeat the same patterns of sound over and over. It is reasonable to suppose that an increase in the variety of human sounds had to precede, or at least coincide with, the evolution of language. Which leads to another question: What makes the variety of sounds coming from a species increase?

As it happens, there is a well-documented case of song variety growing under controlled circumstances. Kazuo Okanoya of the Riken Institute in Tokyo compared songs between two populations of birds: the wild white-rump munia and its domesticated variant, the Bengalese finch. Over several centuries, bird fanciers bred Bengalese finches, selecting them for appearance only. Something odd happened during that time: domesticated finches started singing an extreme and evolving variety of songs, quite unlike the wild munia, which has only a limited number of calls. The wild birds do not expand their vocal range even if they are raised in captivity, so the change was at least in part genetic.

The traditional explanation for such a change is that it must provide an advantage in either survival or sexual selection. In this case, though, the finches were well fed and there were no predators. Meanwhile, breeders, who were influenced only by feather coloration, did the mate selection.

Enter Terry Deacon, a scientist who has made fundamental contributions in widely diverse areas of research. He is a professor of anthropology at the University of California at Berkeley and an expert on the evolution of the brain; he is also interested in the chemical origins of life and the mathematics behind the emergence of complicated structures like language.

Terry offered an unconventional solution to the mystery of Bengalese finch musicality. What if there are certain traits, including song style, that naturally tend to become less constrained from generation to generation but are normally held in check by selection pressures? If the pressures go away, variation should increase rapidly. Terry suggested that the finches developed a wider song variety not because it provided an advantage but merely because in captivity it became possible.

In the wild, songs probably had to be rigid in order for mates to find each other. Birds born with a genetic predilection for musical innovation most likely would have had trouble mating. Once finches experienced the luxury of assured mating (provided they were visually attractive), their song variety exploded.

Brian Ritchie and Simon Kirby of the University of Edinburgh worked with Terry to simulate bird evolution in a computer model, and the idea worked well, at least in a virtual world. Here is yet another example of how science becomes more like storytelling as engineering becomes able to represent some of the machinery of formerly subjective human activities.

Realistic Computationalist Thinking Works Great for Coming Up with Evolutionary Hypotheses

Recent successes using computers to hunt for correlations in giant chunks of text offer a fresh hint that an explosion of variety in song might have been important in human evolution. To see why, compare two popular stories of the beginning of language.

In the first story, a protohuman says his first word for something—maybe *ma* for "mother"—and teaches it to the rest of the tribe. A few generations later, someone comes up with *wa* for "water." Eventually the tribe has enough words to constitute a language.

In the second story, protohumans have become successful enough that more of them are surviving, finding mates, and reproducing. They are making all kinds of weird sounds because evolution allows experimentation to run wild, so long as it doesn't have a negative effect on survival. Meanwhile, the protohumans are doing a lot of things in groups, and their brains start correlating certain distinctive social vocalizations with certain events. Gradually, a large number of approximate words come into use. There is no clear boundary at first between words, phrases, emotional inflection, and any other part of language.

The second story seems more likely to me. Protohumans would have been doing something like what big computers are starting to do now, but with the superior pattern-recognizing capabilities of a brain. While language has become richer over time, it has never become absolutely precise. The ambiguity continues to this day and allows language to grow and change. We are still living out the second story when we come up with new slang, such as "bling" or "LOL."

Even if the second story happened, and is still happening, language has not necessarily become more varied. Rules of speech may have eventually emerged that place restrictions on variety. Maybe those late-arriving rules help us communicate more precisely or just sound sexy and high status, or more likely a little of both. Variety doesn't always have to increase in every way.

Retropolis Redux

Variety could even decrease over time. In Chapter 9, I explained how the lack of stylistic innovation is affecting the human song right now. If you accept that there has been a recent decrease in the stylistic variety, the next question is "Why?" I have already suggested that the answer may be connected with the problem of fragment liberation and the hive mind.

Another explanation, which I also think possible, is that the change since the mid-1980s corresponds with the appearance of

So this is an ironic moment in the history of computer science. We are beginning to succeed at using computers to analyze data without the constraints of rigid grammarlike systems. But when we use computers to create, we are confined to equally rigid 1960s models of how information should be structured. The hope that language would be like a computer program has died. Instead, music has changed to become more like a computer program.

digital editing tools, such as MIDI, for music. Digital tools have more impact on the results than previous tools: if you deviate from the kind of music a digital tool was designed to make, the tool becomes difficult to use. For instance, it's far more common these days for music to have a clockwork-regular beat. This may be largely because some of the most widely used music software becomes awkward to use and can even produce glitches if you vary the tempo much while editing. In predigital days, tools also influenced music, but not nearly as dramatically.

Rendezvous with Rama

In Chapter 2 I argued that the following question can never be asked scientifically: What is the nature of consciousness? No experiment can even show that consciousness exists.

In this chapter, I am wearing a different hat and describing the role computer models play in neuroscience. Do I have to pretend that consciousness doesn't exist at all while I'm wearing this other hat (probably a cap studded with electrodes)?

Here is the way I answer that question: While you can never capture the nature of consciousness, there are ways to get closer and closer to it. For instance, it is possible to ask what meaning is, even if we cannot ask about the experience of meaning.

V. S. Ramachandran, a neuroscientist at the University of California at San Diego and the Salk Institute, has come up with a research program to approach the question of meaning with remarkable concreteness. Like many of the best scientists, Rama (as he is known to his colleagues) is exploring in his work highly complex variants of what made him curious as a child. When he was eleven, he wondered about the digestive system of the Venus flytrap, the carnivorous plant. Are the digestive enzymes in its leaves triggered by proteins, by sugars, or by both? Would saccharin fool the traps the way it fools our taste buds?

Later Rama graduated to studying vision and published his first paper in the journal *Nature* in 1972, when he was twenty. He is best known for work that overlaps with my own interests: using mirrors as a low-tech form of virtual reality to treat phantom-limb pain and stroke paralysis. His research has also sparked a fruitful ongoing dialogue between the two of us about language and meaning.

The brain's cerebral cortex areas are specialized for particular sensory systems, such as vision. There are also overlapping regions between these parts—the cross-modal areas I mentioned earlier in connection with olfaction. Rama is interested in determining how the cross-modal areas of the brain may give rise to a core element of language and meaning: the metaphor.

A Physiological Basis for Metaphor

Rama's canonical example is encapsulated in an experiment known as bouba/kiki. Rama presents test subjects with two words, both of which are pronounceable but meaningless in most languages: bouba and kiki.

Then he shows the subjects two images: one is a spiky, hystricine shape and the other a rounded cloud form. Match the words and the images! Of course, the spiky shape goes with kiki and the cloud matches bouba. This correlation is cross-cultural and appears to be a general truth for all of humankind.

The bouba/kiki experiment isolates one form of linguistic abstraction. "Boubaness" or "kikiness" arises from two stimuli that are otherwise utterly dissimilar: an image formed on the retina versus a sound activated in the cochlea of the ear. Such abstractions seem to be linked to the mental phenomenon of metaphor. For instance, Rama finds that patients who have lesions in a cross-modal brain region called the inferior parietal lobule have difficulty both with the bouba/kiki task and with interpreting proverbs or stories that have nonliteral meanings.

Rama's experiments suggest that some metaphors can be understood as mild forms of synesthesia. In its more severe forms, synesthesia is an intriguing neurological anomaly in which a person's sensory systems are crossed—for example, a color might be perceived as a sound.

What is the connection between the images and the sounds in Rama's experiment? Well, from a mathematical point of view, kiki and the spiky shape both have "sharp" components that are not so pronounced in bouba; similar sharp components are present in the tongue and hand motions needed to make the kiki sound or draw the kiki picture.

Rama suggests that cross-modal abstraction—the ability to make consistent connections across senses—might have initially evolved in lower primates as a better way to grasp branches. Here's how it could have happened: the cross-modal area of the brain might have evolved to link an oblique image hitting the retina (caused by viewing a tilted branch) with an "oblique" sequence of muscle twitches (leading the animal to grab the branch at an angle).

The remapping ability then became coopted for other kinds of abstraction that humans excel in, such as the bouba/kiki metaphor. This is a common phenomenon in evolution: a preexisting structure, slightly modified, takes on parallel yet dissimilar functions.

But Rama also wonders about other kinds of metaphors, ones that don't obviously fall into the bouba/kiki category. In his current favorite example, Shakespeare has Romeo declare Juliet to be "the sun." There is no obvious bouba/kiki–like dynamic that would link a young, female, doomed romantic heroine with a bright orb in the sky, yet the metaphor is immediately clear to anyone who hears it.

Meaning Might Arise from an Artificially Limited Vocabulary

A few years ago, when Rama and I ran into each other at a conference where we were both speaking, I made a simple suggestion to him about how to extend the bouba/kiki idea to Juliet and the sun.

Suppose you had a vocabulary of only one hundred words. (This experience will be familiar if you've ever traveled to a region where you don't speak the language.) In that case, you'd have to use your small vocabulary creatively to get by. Now extend that condition to an extreme. Suppose you had a vocabulary of only four nouns: kiki, bouba, Juliet, and sun. When the choices are reduced, the importance of what might otherwise seem like trivial synesthetic or other elements of commonality is amplified.

Juliet is not spiky, so bouba or the sun, both being rounded, fit better than kiki. (If Juliet were given to angry outbursts of spiky noises, then kiki would be more of a contender, but that's not our girl in this case.) There are a variety of other minor overlaps that make Juliet more sunlike than boubaish.

If a tiny vocabulary has to be stretched to cover a lot of territory, then any difference at all between the qualities of words is practically a world of difference. The brain is so desirous of associations that it will then amplify any tiny potential linkage in order to get a usable one. (There's infinitely more to the metaphor as it appears in the play, of course. Juliet sets like the sun, but when she dies, she doesn't come back like it does. Or maybe the archetype of Juliet always returns, like the sun—a good metaphor breeds itself into a growing community of interacting ideas.)

Likewise, much of the most expressive slang comes from people with limited formal education who are making creative use of the words they know. This is true of pidgin languages, street slang, and so on. The most evocative words are often the most common ones that are used in the widest variety of ways. For example: Yiddish: *Nu?* Spanish: *Pues.*

One reason the metaphor of the sun fascinates me is that it bears on a conflict that has been at the heart of information science since its inception: Can meaning be described compactly and precisely, or is it something that can emerge only in approximate form based on statistical associations between large numbers of components?

Mathematical expressions are compact and precise, and most early computer scientists assumed that at least part of language ought to display those qualities too.

I described above how statistical approaches to tasks like automatic language translation seem to be working better than compact, precise ones. I also argued against the probability of an initial, small, well-defined vocabulary in the evolution of language and in favor of an emergent vocabulary that never became precisely defined.

There is, however, at least one other possibility I didn't describe earlier: vocabulary could be emergent, but there could also be an outside factor that initially makes it difficult for a vocabulary to grow as large as the process of emergence might otherwise encourage.

The bouba/kiki dynamic, along with other similarity-detecting processes in the brain, can be imagined as the basis of the creation of an endless series of metaphors, which could correspond to a boundless vocabulary. But if this explanation is right, the metaphor of the sun might come about only in a situation in which the vocabulary is at least somewhat limited.

Imagine that you had an endless capacity for vocabulary at the same time that you were inventing language. In that case you could make up an arbitrary new word for each new thing you had to say. A compressed vocabulary might engender less lazy, more evocative words.

Maybe the modest brain capacity of early hominids was the source of the limitation of vocabulary size. Whatever the cause, an initially limited vocabulary might be necessary for the emergence of an expressive language. Of course, the vocabulary can always grow later on, once the language has established itself. Modern English has a huge vocabulary.

> If we had infinite brains, capable of using an infinite number of words, those words would mean nothing, because each one would have too specific a usage. Our early hominid ancestors were spared from that problem, but with the coming of the internet, we are in danger of encountering it now. Or, more precisely, we are in danger of pretending with such intensity that we are encountering it that it might as well be true.

Small Brains Might Have Saved Humanity from an Earlier Outbreak of Meaninglessness

If the computing clouds became effectively infinite, there would be a hypothetical danger that all possible interpolations of all possible words—novels, songs, and facial expressions—will cohabit a Borges-like infinite Wikipedia in the ether. Should that come about, all words would become meaningless, and all meaningful expression would become impossible. But, of course, the cloud will never be infinite.

PART FIVE

FUTURE HUMORS

IN THE PREVIOUS SECTIONS, I've argued that when you deny the specialness of personhood, you elicit confused, inferior results from people. On the other hand, I've also argued that computationalism, a philosophical framework that doesn't give people a special place, can be extremely useful in scientific speculations. When we want to understand ourselves on naturalistic terms, we must make use of naturalistic philosophy that accounts for a degree of irreducible complexity, and until someone comes up with another idea, computationalism is the only path we have to do that.

I should also point out that computationalism can be helpful in certain engineering applications. A materialist approach to the human organism is, in fact, essential in some cases in which it isn't necessarily easy to maintain.

For instance, I've worked on surgical simulation tools for many years, and in such instances I try to temporarily adopt a way of thinking about people's bodies as if they were fundamentally no different from animals or sophisticated robots. It isn't work I could do as well without the sense of distance and objectivity.

Unfortunately, we don't have access at this time to a single philosophy that makes sense for all purposes, and we might never find one. Treating people as nothing other than parts of nature is an uninspired basis for designing technologies that embody human aspirations. The inverse error is just as misguided: it's a mistake to treat nature as a person. That is the error that yields confusions like intelligent design.

I've carved out a rough borderline between those situations in which it is beneficial to think of people as "special" and other situations when it isn't.

But I haven't done enough.

It is also important to address the romantic appeal of cybernetic totalism. That appeal is undeniable.

Those who enter into the theater of computationalism are given all the mental solace that is usually associated with traditional religions. These include consolations for metaphysical yearnings, in the form of the race to climb to ever more "meta" or higher-level states of digital representation, and even a colorful eschatology, in the form of the Singularity. And, indeed, through the Singularity a hope of an afterlife is available to the most fervent believers.

Is it conceivable that a new digital humanism could offer romantic visions that are able to compete with this extraordinary spectacle? I have found that humanism provides an even more colorful, heroic, and seductive approach to technology.

This is about aesthetics and emotions, not rational argument. All I can do is tell you how it has been true for me, and hope that you might also find it to be true.

CHAPTER 14

HOMe aT LaST (MY LOVe aFFaIR WITH BaCHeLaRDIan neoTenY)

HERE I PRESENT my own romantic way to think about technology. It includes cephalopod envy, "postsymbolic communication," and an idea of progress that is centered on enriching the depth of communication instead of the acquisition of powers. I believe that these ideas are only a few examples of many more awaiting discovery that will prove to be more seductive than cybernetic totalism.

The Evolutionary Strategy

Neoteny is an evolutionary strategy exhibited to varying degrees in different species, in which the characteristics of early development are drawn out and sustained into an individual organism's chronological age.

For instance, humans exhibit neoteny more than horses. A newborn horse can stand on its own and already possesses many of the other skills of an adult horse. A human baby, by contrast, is more like a fetal horse. It is born without even the most basic abilities of an adult human, such as being able to move about.

Instead, these skills are learned during childhood. We smart mammals get that way by being dumber when we are born than our more instinctual cousins in the animal world. We enter the world essentially as fetuses in air. Neoteny opens a window to the world before our brains can be developed under the sole influence of instinct.

It is sometimes claimed that the level of neoteny in humans is not fixed, that it has been rising over the course of human history. My purpose here isn't to join in a debate about the semantics of nature and nurture. But I think it can certainly be said that neoteny is an immensely useful way of understanding the relationship between change in people and technology, and as with many aspects of our identity, we don't know as much about the genetic component of neoteny as we surely will someday soon.

The phase of life we call "childhood" was greatly expanded in connection with the rise of literacy, because it takes time to learn to read. Illiterate children went to work in the fields as often as they were able, while those who learned to read spent time in an artificial, protected space called the classroom, an extended womb. It has even been claimed that the widespread acceptance of childhood as a familiar phase of human life only occurred in conjunction with the spread of the printing press.

Childhood becomes more innocent, protected, and concentrated with increased affluence. In part this is because there are fewer siblings to compete for the material booty and parental attention. An evolutionary psychologist might also argue that parents are motivated to become more "invested" in a child when there are fewer children to nurture.

With affluence comes extended childhood. It is a common observation that children enter the world of sexuality sooner than they used to, but that is only one side of the coin. Their sexuality also remains child-like for a longer period of time than it used to. The twenties are the new teens, and people in their thirties are often still dating, not having settled on a mate or made a decision about whether to have children or not.

If some infantile trauma or anxiety can be made obsolete by technology, then that will happen as soon as possible (perhaps even sooner!).

Children want attention. Therefore, young adults, in their newly extended childhood, can now perceive themselves to be finally getting enough attention, through social networks and blogs. Lately, the design of online technology has moved from answering this desire for attention to addressing an even earlier developmental stage.

Separation anxiety is assuaged by constant connection. Young people announce every detail of their lives on services like Twitter not to show off, but to avoid the closed door at bedtime, the empty room, the screaming vacuum of an isolated mind.

Been Fast So Long, Feels Like Slow to Me

Accelerating change has practically become a religious belief in Silicon Valley. It often begins to seem to us as though everything is speeding up along with the chips. This can lead many of us to be optimistic about many things that terrify almost everyone else. Technologists such as Ray Kurzweil will argue that accelerating improvement in technological prowess will inevitably outrun problems like global warming and the end of oil. But not every technology-related process speeds up according to Moore's law.

For instance, as I've mentioned earlier, software development doesn't necessarily speed up in sync with improvements in hardware. It often instead slows down as computers get bigger because there are more opportunities for errors in bigger programs. Development becomes slower and more conservative when there is more at stake, and that's what is happening.

For instance, the user interface to search engines is still based on the command line interface, with which the user must construct logical phrases using symbols such as dashes and quotes. That's how personal computers used to be, but it took less than a decade to get from the Apple II to the Macintosh. By contrast, it's been well over a decade since network-based search services appeared, and they are still trapped in the command line era. At this rate, by 2020, we can expect software development to have slowed to a near stasis, like a clock approaching a black hole.

There is another form of slowness related to Moore's law, and it inter-acts with the process of neoteny. Broadly speaking, Moore's law can be expected to accelerate progress in medicine because computers will accelerate the speeds of processes like genomics and drug discovery. That means healthy old age will continue to get healthier and last longer and that the "youthful" phase of life will also be extended. The two go together.

And that means generational shifts in culture and thought will hap-pen less frequently. The baby boom isn't over yet, and the 1960s still pro-vide the dominant reference points in pop culture. This is in part, I believe, because of the phenomena of Retropolis and youthiness, but it is also because the boomers are not merely plentiful and alive but still vigorous and contributing to society. And that is because constantly improving medicine, public health, agriculture, and other fruits of technology have extended the average life span. People live longer as technology improves, so cultural change actually *slows*, because it is tied more to the outgoing generational clock than the incoming one.

So Moore's law makes "generational" cultural change slow down. But that is just the flip side of neoteny. While it is easy to think of neoteny as an emphasis on youthful qualities, which are in essence radical and experimental, when cultural neoteny is pushed to an extreme it implies conservatism, since each generation's perspectives are preserved longer and made more influential as neoteny is extended. Thus, neoteny brings out contradictory qualities in culture.

Silicon Juvenilia

It's worth repeating obvious truths when huge swarms of people are somehow able to remain oblivious. That is why I feel the need to point out the most obvious overall aspect of digital culture: it is comprised of wave after wave of juvenilia.

Some the greatest speculative investments in human history continue to converge on silly Silicon Valley schemes that seem to have been named by Dr. Seuss. On any given day, one might hear of tens or hundreds of millions of dollars flowing to a start-up company named Ublibudly or MeTickly. These are names I just made up, but they would make great venture capital bait if they existed. At these companies one finds rooms full of MIT PhD engineers not seeking cancer cures or sources of safe drinking water for the underdeveloped world but schemes to send little digital pictures of teddy bears and dragons between adult members of social networks. At the end of the road of the pursuit of technological sophistication appears to lie a playhouse in which humankind regresses to nursery school.

It might seem that I am skewering the infantile nature of internet culture, but ridicule is the least of my concerns. True, there's some fun to be had here, but the more important business is relating technological infantilism neoteny to a grand and adventurous trend that characterizes the human species.

And there is truly nothing wrong with that! I am not saying, "The internet is turning us all into children, isn't that awful"; quite the contrary. Cultural neoteny can be wonderful. But it's important to understand the dark side.

Goldingesque Neoteny, Bachelardian Neoteny, and Infantile Neoteny

Everything going on in digital culture, from the ideals of open software to the emergent styles of Wikipedia, can be understood in terms of cultural neoteny. There will usually be both a lovely side and a nasty side to neoteny, and they will correspond to the good and the bad sides of what goes on in any playground.

The division of childhood into good and bad is an admittedly subjective project. One approach to the good side of childhood is celebrated in philosopher Gaston Bachelard's *Poetics of Reverie,* while an aspect of the bad side is described in William Golding's novel *Lord of the Flies.*

The good includes a numinous imagination, unbounded hope, innocence, and sweetness. Childhood is the very essence of magic, optimism, creativity, and open invention of self and the world. It is the heart of tenderness and connection between people, of continuity between generations, of trust, play, and mutuality. It is the time in life when we learn to use our imaginations without the constraints of life lessons.

The bad is more obvious, and includes bullying, voracious irritability, and selfishness.

The net provides copious examples of both aspects of neoteny.

Bachelardian neoteny is found, unannounced, in the occasional MySpace page that communicates the sense of wonder and weirdness that a teen can find in the unfolding world. It also appears in Second Life and gaming environments in which kids discover their expressive capabilities. Honestly, the proportion of banal nonsense to genuine tenderness and wonder is worse online than in the physical world at this time, but the good stuff does exist.

The ugly Goldingesque side of neoteny is as easy to find online as getting wet in the rain—and is described in the sections of this book devoted to trolls and online mob behavior.

My Brush with Bachelardian Neoteny
in the Most Interesting Room in the World

There's almost nothing duller than listening to people talk about inde-
scribable, deeply personal, revelatory experiences: the LSD trip, the
vision on the mountaintop. When you live in the Bay Area, you learn to
carefully avoid those little triggers in a conversation that can bring on the
deluge.

So it is with trepidation that I offer my own version. I am telling my
story because it might help get across a point that is so basic, so ambient,
that it would be otherwise almost impossible to isolate and describe.

Palo Alto in the 1980s was already the capital of Silicon Valley, but
you could still find traces of its former existence as the bucolic border-
lands between the Stanford campus and a vast paradise of sunny
orchards to the south. Just down the main road from Stanford you could
turn onto a dirt path along a creek and find an obscure huddle of stucco
cottages.

Some friends and I had colonized this little enclave, and the atmosphere was "late hippie." I had made some money from video games, and
we were using the proceeds to build VR machines. I remember one day,
amid the colorful mess, one of my colleagues—perhaps Chuck Blan-
chard or Tom Zimmerman—said to me, with a sudden shock, "Do you
realize we're sitting in the most interesting room in the world right
now?"

I'm sure we weren't the only young men at that moment to believe
that what we were doing was the most fascinating thing in the world, but
it still seems to me, all these years later, that the claim was reasonable.
What we were doing was connecting people together in virtual reality for
the first time.

If you had happened upon us, here is what you would have seen. A
number of us would be nursing mad scientist racks filled with computers and an impenetrable mess of cables through whatever crisis of
glitches had most recently threatened to bring the system down. One or
two lucky subjects would be inside virtual reality. From the outside,
you'd have seen these people wearing huge black goggles and gloves
encrusted in patterns of weird small electronic components. Some other

people would be hovering around making sure they didn't walk into walls or trip over cables. But what was most interesting was what the subjects saw from the inside.

On one level, what they saw was absurdly crude images jerking awkwardly around, barely able to regain equilibrium after a quick turn of the head. This was virtual reality's natal condition. But there was a crucial difference, which is that even in the earliest phases of abject crudeness, VR conveyed an amazing new kind of experience in a way that no other media ever had.

It's a disappointment to me that I still have to describe this experience to you in words more than a quarter of a century later. Some derivatives of virtual reality have become commonplace: you can play with avatars and virtual worlds in Second Life and other online services. But it's still very rare to be able to experience what I am about to describe.

So you're in virtual reality. Your brain starts to believe in the virtual world instead of the physical one. There's an uncanny moment when the transition occurs.

Early VR in 1980s had a charm to it that is almost lost today. (I believe it will reappear in the future, though.) The imagery was minimalist, because the computer power necessary to portray a visually rich world did not exist. But our optical design tended to create a saturated and soft effect, instead of the blocky one usually associated with early computer graphics. And we were forced to use our minimal graphic powers very carefully, so there was an enforced elegance to the multihued geometric designs that filled our earliest virtual worlds.

I remember looking at the deeply blue virtual sky and at the first immersive, live virtual hand, a brass-colored cubist sculpture of cylinders and cones, which moved with my thoughts and was me.

We were able to play around with VR as the most basic of basic research, with creativity and openness. These days, it is still, unfortunately, prohibitively expensive to work with full-on VR, so it doesn't happen very much absent a specific application. For instance, before even acquiring equipment, you need special rooms for people to wander around in when they think they're in another world, and the real estate to make those rooms available in a university is not easy to come by.

Full-blown immersive VR is all too often done with a purpose these days. If you are using VR to practice a surgical procedure, you don't have psychedelic clouds in the sky. You might not even have audio, because it is not essential to the task. Ironically, it is getting harder and harder to find examples of the exotic, complete VR experience even as the underlying technology gets cheaper.

It was a self-evident and inviting challenge to attempt to create the most accurate possible virtual bodies, given the crude state of the technology at the time. To do this, we developed full-body suits covered in sensors. A measurement made on the body of someone wearing one of these suits, such as an aspect of the flex of a wrist, would be applied to control a corresponding change in a virtual body. Before long, people were dancing and otherwise goofing around in virtual reality.

Of course, there were bugs. I distinctly remember a wonderful bug that caused my hand to become enormous, like a web of flying skyscrapers. As is often the case, this accident led to an interesting discovery.

It turned out that people could quickly learn to inhabit strange and different bodies and still interact with the virtual world. I became curious about how weird the body could get before the mind would become disoriented. I played around with elongated limb segments and strange limb placements. The most curious experiment involved a virtual lobster. A lobster has a trio of little midriff arms on each side of its body. If physical human bodies sprouted corresponding limbs, we would have measured them with an appropriate bodysuit and that would have been that.

I assume it will not come as a surprise to the reader that the human body does not include these little arms, so the question arose of how to control them. The answer was to extract a little influence from each of many parts of the physical body and merge these data streams into a single control signal for a given joint in the extra lobster limbs. A touch of human elbow twist, a dash of human knee flex; a dozen such movements might be mixed to control the middle joint of little left limb #3. The result was that the principal human elbows and knees could still control their virtual counterparts roughly as before, while also contributing to the control of additional limbs.

Yes, it turns out people can learn to control bodies with extra limbs!

In the future, I fully expect children to turn into molecules and triangles in order to learn about them with a somatic, "gut" feeling. I fully expect morphing to become as important a dating skill as kissing.

There is something extraordinary that you might care to notice when you are in VR, though nothing compels you to: you are no longer aware of your physical body. Your brain has accepted the avatar as your body. The only difference between your body and the rest of the reality you are experiencing is that you already know how to control your body, so it happens automatically and subconsciously.

But actually, because of homuncular flexibility, any part of reality might just as well be a part of your body if you happen to hook up the software elements so that your brain can control it easily. Maybe if you wiggle your toes, the clouds in the sky will wiggle too. Then the clouds would start to feel like part of your body. All the items of experience become more fungible than in the physical world. And this leads to the revelatory experience.

The body and the rest of reality no longer have a prescribed boundary. So what are you at this point? You're floating in there, as a center of experience. You notice you exist, because what else could be going on? I think of VR as a consciousness-noticing machine.

Postsymbolic Communication and Cephalopods

Remember the computer graphics in the movie *Terminator 2* that made it possible for the evil terminator to assume the form and visage of any person it encountered? Morphing—the on-screen transformation—violated the unwritten rules of what was allegedly possible to be seen, and in doing so provided a deep, wrenching pleasure somewhere in the back of the viewer's brain. You could almost feel your neural machinery breaking apart and being glued back together.

Unfortunately, the effect has become a cliché. Nowadays, when you watch a television ad or a science fiction movie, an inner voice says, "Ho hum, just another morph." However, there's a video clip that I often show students and friends to remind them, and myself, of the transportive effects of anatomical transformation. This video is so shocking that most viewers can't process it the first time they see it—so they ask to see it again and again and again, until their mind has expanded enough to take it in.

The video was shot in 1997 by Roger Hanlon while he was scuba diving off Grand Cayman Island. Roger is a researcher at the Marine Biological Laboratory in Woods Hole; his specialty is the study of cephalopods, a family of sea creatures that include octopuses, squids, and cuttlefishes. The video is shot from Roger's point of view as he swims up to examine an unremarkable rock covered in swaying algae.

Suddenly, astonishingly, one-third of the rock and a tangled mass of algae morphs and reveals itself for what it really is: the waving arms of a bright white octopus. Its cover blown, the creature squirts ink at Roger and shoots off into the distance—leaving Roger, and the video viewer, slack-jawed.

The star of this video, *Octopus vulgaris*, is one of a number of cephalopod species capable of morphing, including the mimic octopus and the giant Australian cuttlefish. The trick is so weird that one day I tagged along with Roger on one of his research voyages, just to make sure he wasn't faking it with fancy computer graphics tricks. By then, I was hooked on cephalopods. My friends have had to adjust to my obsession; they've grown accustomed to my effusive rants about these creatures. As far as I'm concerned, cephalopods are the strangest smart creatures on Earth. They offer the best standing example of how truly different intelligent extraterrestrials (if they exist) might be from us, and they taunt us with clues about potential futures for our own species.

The raw brainpower of cephalopods seems to have more potential than the mammalian brain. Cephalopods can do all sorts of things, like think in 3-D and morph, which would be fabulous innate skills in a high-tech future. Tentacle-eye coordination ought to easily be a match for hand-eye coordination. From the point of view of body and brain, cephalopods are primed to evolve into the high-tech-tool-building overlords. By all rights, cephalopods should be running the show and we should be their pets.

What we have that they don't have is neoteny. Our secret weapon is childhood.

Baby cephalopods must make their way on their own from the moment of birth. In fact, some of them have been observed reacting to the world seen through their transparent eggs before they are born, based only on instinct. If people are at one extreme in a spectrum of neoteny, cephalopods are at the other.

Cephalopod males often do not live long after mating. There is no concept of parenting. While individual cephalopods can learn a great deal within a lifetime, they pass on nothing to future generations. Each generation begins afresh, a blank slate, taking in the strange world without guidance other than instincts bred into their genes.

If cephalopods had childhood, surely they would be running the Earth. This can be expressed in an equation, the only one I'll present in this book:

Cephalopods + Childhood = Humans + Virtual Reality

Morphing in cephalopods works somewhat similarly to how it does in computer graphics. Two components are involved: a change in the image or texture visible on a shape's surface, and a change in the underlying shape itself. The "pixels" in the skin of a cephalopod are organs called chromatophores. These can expand and contract quickly, and each is filled with a pigment of a particular color. When a nerve signal causes a red chromatophore to expand, the "pixel" turns red. A pattern of nerve firings causes a shifting image—an animation—to appear on the cephalopod's skin. As for shapes, an octopus can quickly arrange its arms to form a wide variety of forms, such as a fish or a piece of coral, and can even raise welts on its skin to add texture.

Why morph? One reason is camouflage. (The octopus in the video is presumably trying to hide from Roger.) Another is dinner. One of Roger's video clips shows a giant cuttlefish pursuing a crab. The cuttlefish is mostly soft-bodied; the crab is all armor. As the cuttlefish approaches, the medieval-looking crab snaps into a macho posture, waving its sharp claws at its foe's vulnerable body.

The cuttlefish responds with a bizarre and ingenious psychedelic performance. Weird images, luxuriant colors, and successive waves of what look like undulating lightning bolts and filigree swim across its skin. The sight is so unbelievable that even the crab seems disoriented; its menacing gesture is replaced for an instant by another that seems to say,

"Huh?" In that moment the cuttlefish strikes between cracks in the armor. It uses art to hunt!

As a researcher who studies virtual reality, I can tell you exactly what emotion floods through me when I watch cephalopods morph: jealousy.

The problem is that in order to morph in virtual reality, humans must design morph-ready avatars in laborious detail in advance. Our software tools are not yet flexible enough to enable us, in virtual reality, to improvise ourselves into different forms.

In the world of sounds, we can be a little more spontaneous. We can make a wide variety of weird noises through our mouths, spontaneously and as fast as we think. That's why we are able to use language.

But when it comes to visual communication, and other modalities such as smell and spontaneously enacted sculptural shapes that could be felt, we are hamstrung.

We *can* mime—and indeed when I give lectures on cephalopods I like to pretend to be the crab and the cuttlefish to illustrate the tale. (More than one student has pointed out that with my hair as it is, I am looking more and more like a cephalopod as time goes by.) We can learn to draw and paint, or use computer graphics design software, but we cannot generate images at the speed with which we can imagine them.

Suppose we had the ability to morph at will, as fast as we can think. What sort of language might that make possible? Would it be the same old conversation, or would we be able to "say" new things to one another?

For instance, instead of saying, "I'm hungry; let's go crab hunting," you might simulate your own transparency so your friends could see your empty stomach, or you might turn into a video game about crab hunting so you and your compatriots could get in a little practice before the actual hunt.

I call this possibility "postsymbolic communication." It can be a hard idea to think about, but I find it enormously exciting. It would not suggest an annihilation of language as we know it—symbolic communication would continue to exist—but it would give rise to a vivid expansion of meaning.

This is an extraordinary transformation that people might someday experience. We'd then have the option of cutting out the "middleman" of symbols and directly creating shared experience. A fluid kind of concreteness might turn out to be more expressive than abstraction.

In the domain of symbols, you might be able to express a quality like "redness." In postsymbolic communication, you might come across a red bucket. Pull it over your head, and you discover that it is cavernous on the inside. Floating in there is *every* red thing: there are umbrellas, apples, rubies, and droplets of blood. The red within the bucket is not Plato's eternal red. It is concrete. You can see for yourself what the objects have in common. It's a new kind of concreteness that is as expressive as an abstract category.

This is perhaps a dry and academic-sounding example. I also don't want to pretend I understand it completely. Fluid concreteness would be an entirely new expressive domain. It would require new tools, or instruments, so that people could achieve it.

I imagine a virtual saxophone-like instrument in virtual reality with which I can improvise both golden tarantulas and a bucket with all the red things. If I knew how to build it now, I would, but I don't.

I consider it a fundamental unknown whether it is even possible to build such a tool in a way that would actually lift the improviser out of the world of symbols. Even if you used the concept of red in the course of creating the bucket of all red things, you wouldn't have accomplished this goal.

I spend a lot of time on this problem. I am trying to create a new way to make software that escapes the boundaries of preexisting symbol systems. This is my phenotropic project.

The point of the project is to find a way of making software that rejects the idea of the protocol. Instead, each software module must use emergent generic pattern-recognition techniques—similar to the ones I described earlier, which can recognize faces—to connect with other modules. Phenotropic computing could potentially result in a kind of software that is less tangled and unpredictable, since there wouldn't be protocol errors if there weren't any protocols. It would also suggest a path to escaping the prison of predefined, locked-in ontologies like MIDI in human affairs.

The most important thing about postsymbolic communication is that I hope it demonstrates that a humanist softie like me can be as radical and ambitious as any cybernetic totalist in both science and technology, while still believing that people should be considered differently, embodying a special category.

For me, the prospect of an entirely different notion of communication is more thrilling than a construction like the Singularity. Any gadget, even a big one like the Singularity, gets boring after a while. But a deepening of meaning is the most intense potential kind of adventure available to us.

acknowledgments

Some passages in this book are adapted from "Jaron's World," the author's column in *Discover* magazine, and others are adapted from the author's contributions to edge.org, the *Journal of Consciousness Studies*, *Think Magazine*, assorted open letters, and comments submitted to various hearings. They are used here by permission.

Superspecial thanks to early readers of the manuscript: Lee Smolin, Dina Graser, Neal Stephenson, George Dyson, Roger Brent, and Yelena the Porcupine; editors: Jeff Alexander, Marty Asher, and Dan Frank; agents: John Brockman, Katinka Matson, and Max Brockman; at *Discover*: Corey Powell and Bob Guccione Jr.; and various people who tried to help me finish a book over the last few decades: Scott Kim, Kevin Kelly, Bob Prior, Jamie James, my students at UCSF, and untold others.

index

a note about the author

Jaron Lanier is a computer scientist, composer, visual artist, and author. His current appointments include Scholar at Large for Microsoft Corporation and Interdisciplinary Scholar-in-Residence, Center for Entrepreneurship and Technology, University of California at Berkeley.

Lanier's name is also often associated with research into "virtual reality," a term he coined. In the late 1980s he led the team that developed the first implementations of multiperson virtual worlds using head-mounted displays, for both local and wide-area networks, as well as the first "avatars," or representations of users within such systems. While at VPL Research, Inc., he and his colleagues developed the first implementations of virtual reality applications in surgical simulation, vehicle interior prototyping, virtual sets for television production, and assorted other areas. He led the team that developed the first widely used software platform architecture for immersive virtual reality applications. In 2009, he received a Lifetime Career Award from the Institute of Electrical and Electronics Engineers (IEEE) for his contributions to the field.

Lanier received an honorary doctorate from the New Jersey Institute of Technology in 2006, was the recipient of Carnegie Mellon University's Watson Award in 2001, and was a finalist for the first Edge of Computation Award in 2005.

a note on the type

This book was set in Scala, a typeface designed by the Dutch designer
Martin Majoor (b. 1960) in 1988 and released by the FontFont foundry in
1990. While designed as a fully modern family of fonts containing both a
serif and a sans serif alphabet, Scala retains many refinements normally
associated with traditional fonts.

Composed by North Market Street Graphics, Lancaster, Pennsylvania
Printed and bund by RR Donnelley, Harrisonburg, Virginia
Designed by Maggie Hinders